Discoverers of the Universe

Discoverers of

the Universe

William

and

Caroline

Herschel

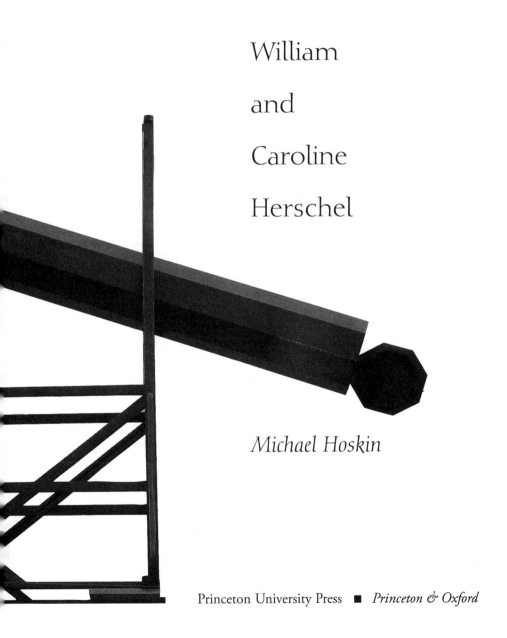

Michael Hoskin

Princeton University Press ■ *Princeton & Oxford*

Copyright © 2011 by Princeton University Press

Published by Princeton University Press, 41 William Street,
   Princeton, New Jersey 08540
In the United Kingdom: Princeton University Press, 6 Oxford Street,
   Woodstock, Oxfordshire OX20 1TW
press.princeton.edu

Jacket art: *Caroline Herschel Runner* from *The Dinner Party*
© Judy Chicago, 1979
Embroidery on linen
Elizabeth A. Sackler Center for Feminist Art
Collection of the Brooklyn Museum
Photo © Donald Woodman/Artists Rights Society (ARS), New York

Library of Congress Cataloging-in-Publication Data
Hoskin, Michael A.
   Discoverers of the universe : William and Caroline Herschel / Michael Hoskin.
      p. cm.
   Includes bibliographical references and index.
   ISBN 978-0-691-14833-5 (cloth : alk. paper)
   1. Herschel, William, Sir, 1738-1822. 2. Herschel, Caroline Lucretia, 1750-
1848. 3. Astronomers—Great Britian—Biography. I. Title.
   QB35.H75   2011
   520.92'241--dc22 [B]   2010031694

British Library Cataloging-in-Publication Data is available

This book has been composed in Adobe Garamond

Printed on acid-free paper ∞

Printed in the United States of America

10 9 8 7 6 5 4 3 2 1

Contents

Illustrations

Figures

Color Plates

Preface

The universe of Newton and Leibniz three centuries ago was mechanical. God was the great clockmaker, and his creation was a vast machinery in which the cogs and wheels cycled endlessly without changing in any significant respect. By contrast, the universe of modern astronomy is biological, in that every feature of it has a life story that we can study. Individual stars are born, develop, mature, and die; at the other extreme, the cosmos itself began with the Big Bang, and we ask what the future holds for it. The man who did more than anyone to bring about this revolution in our conception of the universe was William Herschel; and a subordinate, but nevertheless crucial, contribution to his grand enterprise was made by his sister Caroline. They, if anyone, are the discoverers of the universe in which we see ourselves as living, and this is their story.

It takes only moments to outline the central concept of William's revolutionary thinking. In the later 1780s he was acting the natural historian and assembling specimens of the milky patches in the sky known as "nebulae." Some, at least, of the nebulae were simply clusters of stars, so far away that powerful telescopes were needed to distinguish the individuals. But the very existence of star clusters proved that gravitational attraction (or a similar force) was at work in the heavens, for a cluster must surely have formed because the stars in that region of space had attracted each other and so moved ever closer together.

It was noticeable that in some of the clusters the stars were tightly packed, whereas in others they were widely scattered. Thinking about the implications of this, William realized that as time went on, the outlying stars in a scattered cluster would be pulled inward by the gravitational attraction of the others and so would be drawn more and more toward the center of the cluster. As a result, the cluster would slowly become ever more tightly packed: scattered clusters were young, while tightly packed clusters were old—we live in a biological universe.

Hugely important though this concept is, it represents only a fragment of William's achievement. For half his life he was not an astronomer but a musician, with the ambition to be remembered as a composer. His musical activities matter little to the historian of astronomy; but they mattered to William, and I have tried to give them due prominence.

At that time an astronomer could distinguish himself as a builder of telescopes, or as an observer, or as a theoretician. Unusually—I would argue, uniquely—William excelled in all three. In later life his preeminence as a professional builder of reflecting telescopes went unchallenged throughout Europe; earlier, when he was an amateur astronomer, the unrivalled excellence of the mirror he had ground and polished for his 7-foot reflector was the key to his securing the royal patronage that allowed him to effect the transition from music to astronomy. This mirror enabled him to recognize at a glance that a hitherto unknown planet was no ordinary star. In a trade-off that followed the centuries-old customs of patronage, he named the planet the Georgian Star in honor of King George III, and the king appointed him astronomer to the Court at Windsor. King George would later fund the construction of a monster reflector that was one of the wonders of the age. Meanwhile emperors and kings begged to be allowed to buy a telescope of William's making.

With Caroline as his faithful companion in the night watches, ready to copy down his shouted descriptions, William became one of the great observers of all time. Whereas only a hundred or so nebulae were known before he appeared on the scene, his catalogues contain over two and a half thousand of these mysterious objects and are the forerunners of the New General Catalogue that astronomers use today.

As a theoretician he studied the large-scale universe at a time when almost all contemporary astronomers were preoccupied with our little solar system. And, as we have seen, he portrayed this universe as biological rather than mechanical. He began the transformation of astronomy, from an obsession with the orbits of planets and comets to the cosmology we know today.

Some colleagues of mine have devoted their lives to Isaac Newton and have ended up loathing him as a person and regretting the time spent in his company. By contrast, the Herschels are lovely people, and it has been a privilege to become an invisible member of their family. I hope that when you finish this book, you will feel the same.

I owe three major debts. The first is to John Herschel-Shorland, William's great-great grandson, whose cooperation no historian has ever sought in vain. The second is to Anthony Turner, who selflessly turned over to me the store of Herschel materials he had assembled. The third is to the Royal Astronomical Society, owners of the major archive of Herschel manuscripts. Forty years ago the RAS paid to have the archive sorted and catalogued. They then had the papers microfilmed and made available for purchase by institutions worldwide. And then, when new technology became available, the microfilms were put on CDs, so that now anyone interested can purchase the entire collection for no more than the price of a good dinner. For a historian, to have this vast treasure trove immediately available at the touch of one's keyboard is a privilege beyond imagining.

In conclusion I thank Ingrid Gnerlich, Beth Clevenger, and Gail Schmitt of Princeton University Press, who combine efficiency with charm. My collaboration with them has been a delight.

The Herschel Family

Isaac and Anna Herschel

Isaac was born Jan. 14 [*not* 4], 1707, Hohenziaz (now part of Magdeburg). Married ANNA ILSE [or ILSA] MORITZEN, Oct. 12, 1732, Schlosskirche, Hanover. Died Mar. 22, 1767, Hanover; buried in Gartenfriedhof. Anna was born late Dec. 1712 or early Jan. 1713, Neustadt am Rübenberge. Died Nov. 19, 1789, Hanover; buried in same grave as Isaac (and, later, Caroline). Ten children, as follows.

Sophia Elizabeth Griesbach [Sophia; née Herschel]

Born Apr. 12, 1733, Hanover. Married bandsman JOACHIM HEINRICH GRIESBACH, Jan. 21, 1755, Hanover. Died March 30, 1803, Hanover (buried Apr. 3). Heinrich was born July 23, 1730, Bodenwerder; died Jan. 31, 1773, Coppenbrügge (buried Feb. 3). Seven children:

GEORGE LUDOLPH JACOB GRIESBACH [GEORGE], born Oct. 11, 1757, Hanover (baptized Oct. 13). Married his pupil MARY ANN WRIGHT SMITH, Oct. 31, 1786, Windsor; nine children. Died Nov. 28, 1824, Windsor. Mary died Jan. 27, 1815, Windsor.

CARL FRIEDRICH LUDWIG GRIESBACH [CHARLES], born Mar. 5, 1760, Coppenbrügge. Married SARAH WIGG, Dec. 15, 1796, Medstead; ten children. Died Mar. 20, 1835, Pocklington, Yorks. Sarah was born on July 8, 1776, and died 1830 at Baden near Vienna, presumably on a tour.

JUSTUS HEINRICH CHRISTIAN GRIESBACH [HENRY], born Apr. 22, 1762, Coppenbrügge. Married (i) MARY BLAKENEY, June 27, 1791; five children. Mary was born June 27, 1764, and died Jan. 25, 1831; (ii) MARY BLOEMFIELD, Oct. 10, 1832. Henry died Feb. 9, 1833, Windsor.

ANNE ELEONORE CHARLOTTE GRIESBACH, born Feb. 17, 1764, Coppenbrügge. Died there of smallpox, Dec. 17, 1766.

FRIDERICA WILHELMINA GRIESBACH, born Feb. 24, 1767, Coppen-brügge. Still living in Germany (unmarried) in 1825.

JOHANN FRIEDRICH ALEXANDER GRIESBACH [FREDERICK], born June 2, 1769, Coppenbrügge. Married MARY FRANCES WYBOROW, daughter of the first master cook at Windsor Castle, Oct. 16, 1792; eight children. Died Jan. 11, 1825, at Putney or Brompton.

JOHANN WILHELM GRIESBACH [WILLIAM], born Jan. 10, 1772, Coppenbrügge. Unmarried, one daughter. Died May or June 1825.

Heinrich Anton Jacob Herschel [Jacob]

Born Nov. 20, 1734, Hanover. Unmarried. Died (by strangulation), June 23, 1792, Hanover.

Johann Heinrich Herschel

Born Apr. 25, 1736, Hanover. Died Nov. 26, 1743, Hanover.

Friedrich Wilhelm Herschel [William]

Born Nov. 15, 1738, Hanover. Married MARY PITT, widow, May 8, 1788, Upton Church. Died Aug. 25, 1822, Slough (buried in Upton Church). Mary was born June 13, 1750, the daughter of ADEE BALDWIN (1717–69; son of THOMAS BALDWIN and GRACE ADEE) and his wife ELIZABETH BROOKER (died Oct. 22, 1798). Mary's only sibling to survive infancy, THOMAS BALDWIN (1754–1821), married MARTHA PARSONS, 1786; three children: SOPHIA BALDWIN (May 21, 1783–Nov. 2, 1820), who married THOMAS BECKWITH, Mar. 11, 1819 (one son, 1820–21); MARY BALDWIN (1788–1873); and THOMAS BALDWIN (1791–1853), who married MARY ROSE. The future Mary Herschel's first marriage was to JOHN PITT (1723–86); two children, PAUL ADEE PITT (born 1773, died Feb. 1793) and WILLIAM (1783). Mary died Jan. 6, 1832, Slough (buried in Upton Church). One child:

JOHN FREDERICK WILLIAM HERSCHEL [JOHN], born Mar. 7, 1792, Slough. Married MARGARET BRODIE STUART [Maggie], Mar. 3, 1829, St Marylebone, London; twelve children. Died May 11, 1871, Hawkhurst, Kent (buried next to Newton in Westminster Abbey). Maggie was born Aug. 16, 1810; died Aug. 3, 1884.

Anna Christina Herschel

Born July 12 or 13, 1741, Hanover. Died of whooping cough, July 22, 1748, Hanover.

Johann Alexander Herschel [Alexander]

Born Nov. 13, 1745, Hanover. Married Mrs. MARGARET SMITH, July 31, 1783, Walcot Church, Bath. No children. Died, Mar. 16, 1821, Hanover (buried in Gartenfriedhof, Mar. 20). Margaret died, c. Feb. 5, 1788, Bath (buried at Weston, Bath, Feb. 10).

Maria Dorethea Herschel

Born June 8, 1748, Hanover. Died Apr. 21, 1749, Hanover.

Carolina Lucretia Herschel [Caroline]

Born Mar. 16, 1750, Hanover. Died Jan. 9, 1848, Hanover (buried in vault over parents' grave in Gartenfriedhof, Jan. 18).

Frantz Johann Herschel

Born, May 13, 1752, Hanover. Died (of smallpox), Mar. 26, 1754.

Johann Dietrich Herschel [Dietrich]

Born Sept. 13, 1755, Hanover. Married CATHARINA MARIA REIFF, Oct. 5, 1779, Schlosskirche, Hanover. Died Jan. 19, 1827, Hanover (buried in Neustaedter Hof- und Stadtkirche). Catharina was born Jan. 16, 1760, Hanover, and died Dec. 1, 1846, Hanover (buried in Schlosskirche); she was the second daughter of Georg Heinrich Reiff (1721–Dec. 30, 1804) and Anna Elizabeth, née Lindemann (1717–Nov. 5, 1788). Four children:

GEORG HEINRICH HERSCHEL, born Mar. 7, 1781, Hanover (baptized Mar. 15). Died of yellow fever, 1806, Charlestown, Mass., USA.

ANNA ELIZABETH HERSCHEL, born June 17, 1783, Hanover (baptized June 22). Married CHRISTIAN PHILIPP KNIPPING, July 30, 1802. Nine children. Died Feb. 4, 1872, Hanover. Christian was born Hemeringen, Jan. 21, 1760; died July 2, 1822, Lachen.

Sophia Dorothea Herschel, born June 3, 1785, Hanover (baptized June 12). Married Dr. Johann Friedrich Wilhelm Richter, Jan. 3, 1809. Four children. Died Jan. 12, 1861. Johann died Feb. 1832.

Caroline Wilhelmine Marie Antonie Herschel, born June 10, 1799, Hanover (baptized June 12), living there in June 1864. Married Dr. David Groskopf; one son.

Discoverers of the Universe

AUGUST 1772
The Partnership Convenes

On Monday, August 24, 1772, after two days and two nights of misery, the seasick passengers on the packet boat from Holland at last reached dry land on England's east coast.[1] Out to sea, their ship lay at anchor, half-wrecked, its mainmast broken by the storm. The passengers had had to be "thrown like balls" on the shore by burly sailors from a small open boat.

One of them was the twenty-two-year-old Carolina Lucretia Herschel (figure 1) from Hanover in Germany. Carolina, or Caroline as she became known, was unprepossessing—tiny, well under five feet in height, and her face pockmarked by the smallpox she had suffered as a child. As Isaac, her loving but blunt father, had foreseen, she would die "a poor solitary old maid."[2]

Another passenger was Caroline's brother Friedrich Wilhelm, her senior by a dozen years. William, to give him the name he had long since adopted in England and that would become formally his by Act of Parliament in 1793,[3] had made this journey several times, the first as a boy bandsman in the Hanoverian Guards when they were summoned in 1756 to the defense of Britain against the French. Caroline, by contrast, had rarely ventured outside Hanover, and she was full of apprehension as to what lay ahead. She was on the windswept beach because William—Prince Charming to her Cinderella[4]—had traveled to Hanover to rescue her from her life of relentless drudgery as a slave to their mother. Anna had seen her daughter as a lifelong source of cheap and reliable domestic help, and so she had ruthlessly prevented Caroline from learning the skills required to get a job outside the family home as a seamstress, a governess of children, a teacher of music. Governesses were expected to teach French to their charges, and so Anna simply refused to allow Caroline to learn the language. The Herschels were a talented musical family, and all the boys had become professional musicians; but Isaac, now sadly long dead, had been allowed to give his daughter the occasional violin lesson only when Anna was in a particularly good mood, or safely out of the house. It had taken imagination and cunning for William to pry Caroline from her mother's clutches.

William was by now a leading figure in the musical life of Bath, the fashionable West of England town where the aristocracy came to take the waters during the winter season. All three of his surviving brothers—Jacob, Alexander and Dietrich—had spent time with him in Bath; and so rich were the musical pickings there that Alexander had quit his prestigious position in the Hanoverian Court Orchestra to make Bath his home. In Hanover, if the musicians of the court orchestra undertook outside engagements, it had to be without pay; in Bath, things were very different.

In the autumn of 1771 William and Alexander had hatched a plot to rescue Caroline. Alexander had heard Caroline sing and thought her untrained voice showed promise. William had therefore written to their mother proposing that Caroline come to Bath for a trial period of a couple of years to see whether her voice was good enough for her to perform in events such as the Handel oratorios that William staged during Holy Week. If it was not, he would send her home. To imagine that this diminutive, disfigured, German, working-class girl might stand in front of an aristocratic Bath audience and sing "I Know That My Redeemer Liveth" was preposterous; but it was the best excuse they could think of. Jacob, their eldest brother and now head of the family back in Hanover, had ridiculed the idea. But by a lucky chance he had been out of town with the court orchestra, entertaining the Queen of Denmark in a hunting lodge, when William arrived on August 2, 1772, to press the matter to a conclusion. Anna was won round by the promise of an annuity to pay for substitute help, which offered a permanent solution to her domestic problems. But Caroline hesitated at the daunting choice before her. "I tried to still the compunction I felt at leaving relatives who I feared would lose some of their comforts by my desertion."

William had engagements in Bath, and he could not afford to linger in Hanover for more than a fortnight. Caroline was a grown woman, but she was most reluctant to leave without Jacob's permission. Letters were slow to travel the eighty miles between Hanover and the hunting lodge, and eventually Caroline was forced to make up her own mind: was it to be Hanover with the ruthless Anna and the domineering Jacob, or life in a foreign land with her beloved William? In the first of the four major

Figure 1. The only known image of Caroline as a young woman, painted before she left Hanover in 1772 for Bath. Even poor families might be able to afford such a "shade" (silhouette), and with care completely faithful copies could be made. The shade evidently remained in Hanover with the Herschels, who by 1803 were reduced to Dietrich Herschel and his family. Caroline returned to Hanover in 1822 and became a local celebrity, and it may be that several copies were then made. This one was sent to the leading English amateur astronomer, John Lee, by Dr. Georg Friedrich Grotefend of Hanover in 1844, and two (one perhaps the original) were sent in 1864 from Dietrich's youngest daughter to Caroline's nephew John in England. Courtesy of the Museum of the History of Science, Oxford University.

decisions of her life, she opted for Bath, even though "I was at last obliged to part from my dear Mother and most dear Dietrich . . . without taking the consent to my going from my eldest brother along with me." And so, on the sixteenth, William and Caroline took the Postwagen from Hanover. They spent six stormy days and nights in the open air before they reached the Dutch coast. Still the storm raged, and they had to be rowed two miles in an open dinghy before they reached the packet boat. And so it was that the seasick and bedraggled Caroline found herself on the beach of a land of whose language she spoke not a word.

She and William made their uncertain way to a nearby house, where they found some of their ravenous fellow passengers already breakfasting from fine wheaten loaves that serving women were cutting "as fast as ever they could." After they had eaten, one of the women took Caroline upstairs to change her clothes. Fortified, she and William climbed into a cart that was to take them to a stopping place where they could await a public stagecoach (a diligence) bound for London. But the horse pulling the cart was not accustomed to working between shafts, and it bolted, overturning the cart. Caroline and William were thrown out, Caroline ending up in a ditch that fortunately proved to be dry. They each checked their limbs and found them intact. Luckily, a gentleman on horseback was nearby with his servant, and he came to their rescue. A true Samaritan, he escorted them safely all the way to an inn in the city of London.

William had business to transact, and so he left Caroline in the inn with their fellow passengers, but he returned in the evening and invited her to join him in a visit to opticians' shops. This naturally puzzled Caroline at the time. Only later did it dawn on her that William was displaying early symptoms of his obsession with astronomy. The constellations that he had introduced her to while they were traveling by coach across Holland had been a first step toward coaxing her into sharing his passion.

Caroline's hat was in one of the Dutch canals, so she persuaded the landlady of the inn to lend her one of her daughter's. The next day William took her to see St. Paul's Cathedral, the Bank of England, and other sights, and in the evening they made their way to the inn from which the Bath coach departed at ten o'clock. They may well have stopped for breakfast—or at least to change horses—at the Crown Inn in the tiny village of Slough, a couple of miles north of Windsor Castle. They may even have met a young woman of Caroline's age whose family owned the inn; if so, William had glimpsed his future wife.

Caroline grew increasingly anxious as they neared Bath, where they arrived at four in the afternoon. She was "in a strange Country and among straingers"—but at least Alexander would be there with a greeting in German for her. No such luck. Alexander was in Southampton, for Bath was now out of season and its musicians were looking for work elsewhere, often enough in stately homes where they would coach the ladies of the house-hold and entertain guests after dinner. But Alexander's absence hardly mattered, for Caroline was "almost annihilated," not having slept properly for nearly a fortnight. She drank some tea, went to bed, and did not awake until the following afternoon.

When Alexander returned to Bath, the three members were assembled of what would become the greatest partnership astronomy has yet known.

1

1707–1773
A Musician's Odyssey

William's early life in Hanover had not been easy. Despite Isaac's absences for military service, he succeeded in fathering ten children; of the ten, four sons and two daughters survived into adulthood.[1] Isaac was a poorly paid bandsman in the Hanoverian Guards; but he did what he could to supplement the basic education offered his children by the Garrison School, which consisted of reading, writing, and religious knowledge to the age of fourteen for both boys and girls, and for the boys, arithmetic. William later recorded how "my father's great attachment to Music determined him to endeavour to make all his sons complete Musicians." As soon as the boys were old enough to hold a miniature violin, their lessons would begin.

Isaac, Anna, and the Founding of the Herschel Dynasty

Isaac himself was born in 1707, the youngest son of Abraham Herschel, a gardener who worked on an estate near Magdeburg, midway between Berlin and Hanover. Abraham was a remarkable man, for he "was very fond of the art of arithmetic and writing as well as of drawing and music," and when he returned home after a hard day's manual work he would wash his hands, eat his supper, and then stretch his mind with pen and paper. Unfortunately Abraham died when Isaac was only eleven, and his widow could not afford to put her son through the usual apprenticeship as a gardener. But the resourceful boy taught himself the rudiments of gardening and eventually got a job tending the garden of an aristocratic widow. Yet he found himself irresistibly drawn to music. As a lad he had managed to buy a violin, and he had taught himself to play by ear. Now he used his wages to purchase an oboe and to pay for proper lessons.

When he was twenty-one, Isaac took his courage in both hands, quit his gardening job, and set off for Berlin to find a post as oboist, only to decide that what he was offered was "very bad and slavish." Impressed by the young man's dedication, his surviving brother and their sister paid for him to have a year's musical tuition with an elderly Prussian band conductor. Eventually we find him in Brunswick, again looking for a post as oboist—and again declining what he was offered, this time because it was "too Prussian." Isaac next traveled to Hanover, where the elector—who was also King of Britain—maintained a corps of Guards with its own band. This time Isaac found the terms acceptable, and on August 7, 1731, he at last became a professional musician.

Isaac was a young man far from home. Hanover was a prosperous city whose houses employed servant girls from the surrounding countryside, and among them was Anna Ilse Moritzen, the illiterate eighteen-year-old daughter of a baker. For the first time in her life she was free from family constraints. She met the lonely bandsman, they went to bed, and conceived a daughter. Normally weddings took place in the bride's village and were accompanied by great festivities, but Anna's pregnancy made this impossible. Instead they were married quietly on October 12, 1732, in the Garrison Church in Hanover.[2] Six months later, on April 12, 1833, Sophia was born.

Thus was founded the great Herschel dynasty. No fewer than ten of their immediate descendants would be at one time or another in the service of King George III or his consort, Queen Charlotte: their sons Jacob, Alexander, and Dietrich would be members of the elector's court orchestra in Hanover; Sophia's five sons would form the core of Queen Charlotte's band at Windsor Castle; while William, and later Caroline, would become salaried astronomers to the Court at Windsor. And William's son John would be awarded the hereditary title of baronet by Queen Victoria for his services to astronomy, and when he died he would be buried in Westminster Abbey, next to Newton. As an educator of children, Isaac was without peer.

Trials and Tribulations of an Army Bandsman

In times of peace the routine of a bandsman in the Guards had much to commend it—minimal duties and maximal family life. By 1741 Sophia had three brothers—Jacob, William, and Johann Heinrich, who was to die young—and a sister, Anna Christina, who would also die in childhood. But now the War of the Austrian Succession was raging, and in wartime the bandsmen would go on campaign along with the fighting soldiers, separated from their families and enduring hardships and privation. That September the Guards marched out of Hanover, only to return six weeks later. The following September they marched again, and in June 1743 they fought in the Battle of Dettingen. Although the Hanoverians were victorious, Isaac and his comrades spent the following night in a waterlogged field. For a time Isaac lost the use of his limbs, and his health would never recover. After his convalescence he was granted a spell of home leave, during which the fertile Anna conceived Alexander.

In February 1746 the Guards returned home. Isaac had had his fill of army life, and he applied for "dismission" (discharge). But how was he to earn his living? There were churches aplenty in Hanover, but he was no organist. The court orchestra offered prestige and affluence and entry into the higher echelons of society, but vacancies were rare, and in any case a humble gardener-turned-bandsman could hardly aim so high. And so when winter came Isaac decided to transfer his family to the great port of Hamburg, where surely there would be demand for musicians.

The journey was difficult—Alexander was just one year old, his milk bottle was frozen, and the vigor of his complaints would live in the memory of those who had to endure them—and on arrival Hamburg proved to be populated by philistines unconcerned about music. While Isaac pondered what to do next, he chanced to meet a former pupil, General Georg August von Wangenheim, no less. The general made Isaac an offer he could not refuse. The prospects of peace were good, the general said, and if Isaac returned to Hanover he could rejoin the Guards band, confident that he could live at home in peace for many years to come. Not only that, but Isaac's talented eldest son, Jacob, could join his father in the band, and William might do the same when he attained the age of fourteen, the age when a schoolchild would be confirmed and go out into the world.

And so the Herschels returned to Hanover. Isaac rejoined the Guards, Jacob marched in the band alongside his father, and they and their comrades were able to live at home in peace and contentment. In May 1753 William, now aged fourteen, was auditioned on the oboe and violin by General Sommerfeld, and so a third Herschel joined the band. Isaac, despite his continuing health problems, was enjoying the happiest time of his life. He could supervise his three sons as they developed their talents as musicians, and the wages earned by Jacob and William could be used to further their general education: lessons in French for both, and for the enquiring William an introduction to logic, ethics, and metaphysics. Isaac was an exemplary father.

But soon war clouds began to gather once more. The French were, as ever, on the march, and in the spring of 1756 the Hanoverian Guards were summoned to England by the current Elector of Hanover to reinforce the realm he ruled as King George II. This early move in what we know as the Seven Years' War proved to be a false alarm, but William took the opportunity to learn some English. He saved up enough pennies to buy John Locke's great three-volume treatise, *An Essay Concerning Human Understanding*, but what his fellow bandsmen thought of this pretentious acquisition, we are not told. Both boys made friends among the musical community of London and the surrounding area, and this would one day stand them in good stead.

Jacob had long since decided that the army was not for him, and such were his musical talents that he had been assured that a post in the court orchestra would be his just as soon as formal approval reached Hanover from the elector's entourage in London. To his chagrin, the required letter had still not arrived when the Guards—with Jacob resentfully among them—left for England. When the letter did arrive Jacob was in England and so unavailable, and the post went to another candidate. But at least he now succeeded in securing his dismission from the army. Comfort was always near the top of his priorities, and in the autumn he returned home by ship and coach, followed in January by Isaac and William, who had to march across Germany with their regiment while the bandsmen did their best to lift the spirits of the troops.

Defeat at Hastenbeck

Before long, sadly, the French were once again on the warpath, and this time the threat was real and against Hanover itself. In July 1757 the Hanoverian Guards were defeated at Hastenbeck, some twenty-odd miles from Hanover. As musicians, Isaac and William were entitled to take cover when shooting began. If we are to believe what William told his son in later life, while the battle raged, "with balls flying over his head he walked behind a hedge spouting speeches, rhetoric then being his favourite study."[3]

In the chaotic aftermath of Hastenbeck, Isaac persuaded himself that because William was a boy and not under oath, he was free to quit the field of battle, and so Isaac discreetly sent him home to Hanover, where Jacob was lying low.[4] But William found that the burghers of Hanover were desperately trying to raise a makeshift force to defend the city against the French, and all able-bodied men were being pressed into service. Suddenly civilian life lost its attraction: back in the army, as a bandsman, William was accepted by both sides as a noncombatant. So he borrowed a civilian greatcoat from the family's landlord, and wearing it, slipped out of the city, followed at a discreet distance by his mother carrying a bundle containing his uniform. Once outside the ring of pickets, he changed into uniform and bade Anna farewell. Before long he was back with the band, relieved to find that his absence had not been noticed.

Isaac was far from pleased at the reappearance of his son, whom he had supposed to be out of harm's way. The following weeks of retreat and confusion were arduous and fraught for soldiers and bandsmen alike, and finally Isaac uttered to William in German the equivalent of "Why don't you get the hell out of here?" England offered sanctuary, and a musician could earn a living in any land. So Isaac dispatched him to the port of Hamburg, there to await the arrival of Jacob, to whom Isaac somehow got word of his plan for the two boys' removal to safety. Their travel he paid for by borrowing to the limit from one of his pupils; but on arrival in England the boys would have to fend for themselves.

William had managed to get word to his mother asking her to forward his possessions to Hamburg, including some celestial globes he had made himself. But the illiterate woman had little patience with such trinkets; she instead gave them to Caroline and her baby brother, Dietrich, as playthings, and before long they were in pieces.

William the Refugee Deserter

The boys arrived penniless—William had a single French crown piece in his pocket—and for two years they survived in and around London by copying music and getting whatever teaching and performing engagements they could. Matters were not helped by the talented Jacob's absolute refusal to play second fiddle: if he was asked to be part of a band, he must be the first violin, failing which he would decline to accept such "degradation," even if it meant going hungry. At last, in the autumn of 1759, the

French were expelled from Hanover, and Jacob returned home. But he traveled alone: William was formally a deserter (Isaac had been arrested briefly for conniving at his escape) and had no wish whatever to be compelled to rejoin the army.[5]

But London was overstocked with musicians, and so when early the following year William was invited to take charge of a small military band in the north of England, he accepted with alacrity. He had given Jacob every penny he could spare to help pay for his brother's travel home, but he made himself solvent again by walking the lengthy journey to Richmond in Yorkshire and pocketing his traveling expenses.

The post with the band was part-time, but it gave William a secure base from which to branch out as teacher, performer, and especially composer of music (figure 2). It was as a composer that he hoped one day to be remembered, and in his methodical fashion he was soon turning out symphonies at the rate of half a dozen a year.[6] He also began to write Jacob formal "letters" on philosophy and music and suchlike. He hoped that one day these minitreatises would be published. They are priggish and pretentious, but they reveal an original mind restless under the constraints of the daily round of the itinerant musician.

In the Garrison School back in Hanover, William had been an able pupil who helped the master by supervising the lessons of the younger children. At home, Isaac had done all in his power to further his sons' education. Although he never had money to spare for books, Isaac somehow managed to teach himself something of the ideas of the great mathematicians, and he encouraged his sons to see these ideas, not as received wisdom but as claims to be debated. Caroline recorded:

> But generally their conversation would branch out on Philosophical subjects, when my brother W^m and his Father often were arguing with such warmth and my Mother's interference became necessary when the names Leibnitz, Newton and Euler sounded rather too loud for the repose of her little ones; who ought to be in school by seven in the morning. But it seems that on the brothers' retiring to their own room, where they partook of one bed, my brother W^m still had a great deal to say; and frequently it happened that when he stopt for an assent or reply; he found his hearer had gone to sleep, and I suppose it was not till then that he bethought himself to do the same.

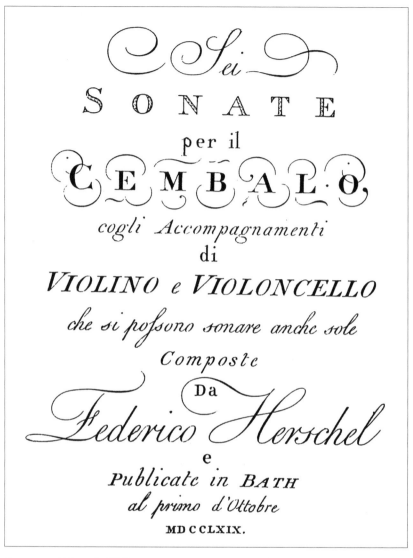

Figure 2. Although William wrote numerous symphonies and concertos while in the north of England, the only substantial musical composition he published in his lifetime was this set of six sonatas for harpsichord with optional violin and cello accompaniment, which date from his early years in Bath. The violin and cello parts are slight and contribute little, but the harpsichord part calls for a level of virtuosity and is designed for an instrument with a swell mechanism. William describes the sonatas as "Lessons for Scholars" (RAS W.7/11, 17), but if so he was blessed with talented pupils.

In Sunderland William at last had time to indulge his fascination for mathematics. As a musician he was intrigued by the arithmetic underlying musical harmony, and so he bought a copy of *Harmonics* by the Cambridge professor Robert Smith.[7] He liked it so well that in the early 1770s, when he came across Smith's other work, the two-volume *Opticks*,[8] he decided to buy that too, with momentous consequences for the history of astronomy.

But this was for the future, and meanwhile he had a living to earn. Early in 1761 he heard of a vacancy that could bring him both prestige and money: that of manager of the concerts in the Scottish capital city of Edinburgh. The incumbent, he was told, intended to resign. William journeyed north and was delighted to be introduced to the great philosopher David Hume. A few days later he was invited to lead a local band in concerts that included some of his own music.

> Mr. Hume, who patronised my performance, asked me to dine with him and accepting of his invitation I met a considerable company, all of whom were pleased to express their approbation of my musical talents.[9]

He returned south confident that the post was his, and so he resigned from the military band.

Alas, the manager in Edinburgh changed his mind, and now William was without the security of the regular source of income he had enjoyed until now. He was sure he could make a living from his freelance work, but, as he told Jacob, "a certain anxiety attends a vagrant life. I do daily meet with vexations and trouble and live only by hope."[10] He was constantly journeying on horseback in all weather. If it was a sunny day, he would pass the time by reading a book while the horse made its own way forward. On one occasion the horse took exception to something and reared up, after which William found himself on the ground facing the horse, with the book still in his hand.[11] But in winter he would have to brave the elements:

> I will only say that at 9 o'clock, when I had still about 20 miles to ride, I was caught in an unusually heavy thunderstorm, which continued accompanied by torrents of rain, with unbroken fury, for three hours, and threatened me with sudden death. The distance from an habitation, the darkness and loneliness, obliged me nevertheless to ride on. I pursued my way therefore with unshaken sang-froid although I was often obliged to shut my eyes on account of

the blinding lightening. At last the flashes all around me were so terrifying that my horse refused to go on; luckily at this moment I found myself near a house, into which, after much knocking, I was admitted. This morning, at 3 o'clock, I proceeded on my journey and arrived safely at this place.[12]

Things got so bad that on one occasion he confessed to Jacob,

I have for some time been thinking of leaving off professing Musick and the first opportunity that offers I shall really do so. It is very well, in your way, when one has a fixed Salary, but to take so much for a Concert, so much for teaching, and so much for a Benefit is what I do not like at all.[13]

In the spring of 1762 came the breakthrough. William took part in a concert in Leeds at which one of his own symphonies was played to considerable acclaim, and five days later there was another concert during which the audience insisted that William and a visiting violinist play the same piece in head-to-head competition. So impressed with William were the burghers of Leeds that in no time at all he found himself appointed the director of concerts.

Entries in the *Leedes Intelligencer* record the ups and downs of musical life in the town. As elsewhere, musical events were either organized by the director and funded from the sale of tickets for the entire series (in which case the quality of what could be offered would depend upon the number of subscriptions taken up), or "benefit" concerts privately promoted in the hope of making a profit—an inevitable source of rivalry among musicians of the town. William figures constantly in the *Intelligencer*, as performer and/or composer, and of course as director of concerts. For example, on April 12, 1763, the *Intelligencer* announced:

Mr Herschel

Takes this Opportunity of returning Thanks to his Friends for the great and many Favours he has met with, since he came to Leedes, and is particularly oblig'd to them for their kind Approbation of his Conduct at the Subscription Concert.

He hopes to have the Continuance of their Kindness and flatters himself that those superior connoisseurs who have discovered any imperfections in the Musical Part of the Concert will easily excuse them when they reflect on the Cheapness of the Terms and Number of the Concerts; at the same Time believes that those Imperfections might easily be remedied another Season, by some small alteration in the Proposals.

He also takes the Liberty of acquainting them, that he intends to teach the *Harpsichord*, the *Guitar*, and the *Violin*, to the best of his capacity, and on the most reasonable Terms.

This declaration was repeated the following week, which provoked William's chief rival, a Mr. J. Crompton, to make a riposte. Readers, he said, should not interpret William's offer of lessons to imply that he, Crompton, had given up teaching. To the contrary; and *his* pupils did not need to send to London for instruments and musical scores, for he himself kept a stock of these for their use. Not only that, but he had harpsichords, spinets, guitars, and English harps for sale, "as good as new."

In March 1762 Jacob, a member of the Hanoverian Court Orchestra, used his considerable influence to secure his brother's formal dismission from the Hanoverian Guards. The printed document (figure 3) pays the standard tribute to soldiers and bandsmen seeking honorable discharge: William had "borne himself on all warlike occasions, marches and guards right manfully as becomes a good hautboist."[14] He was no longer a deserter; but his period in the shadows would always embarrass him, so much so that when in later life he was asked about the circumstances of his arrival in England he would be "economical with the truth," as the saying goes.[15] However, he was now free to visit Hanover, and in 1764 he arranged a visit to reassure his anxious father as to his prospects in England. Indeed, Isaac, who was in failing health, cherished hopes of persuading his son to return home.

William's Visit to Hanover

On April 1, 1764, the day before William (plate 1) arrived, there was an eclipse of the sun. William observed it from his coach as it crossed Lüneburg Heath. In Hanover, Isaac gathered his family around a water butt in a courtyard so that they could safely view the sun in the reflection, while he explained how such an eclipse occurs.

At noon the following day, William reached home, putting his family

Figure 3. William's formal discharge from the Hanoverian Guards. Photograph by the late W. H. Steavenson, courtesy of the Institute of Astronomy, Cambridge University.

into what Caroline described as "a Tumult of joy." Before long a young gentleman arrived pretending to have come for a lesson with Isaac, but he was soon unmasked as William's brother Alexander, who had been eleven years old when they had last seen each other, during William's clandestine return home in the aftermath of the Battle of Hastenbeck.

The intervening years had not been kind to Alexander.[16] Isaac had been interned along with the rest of the Hanoverian troops, and so he had been unable to give his young son the musical instruction essential for his future career. With many misgivings, Isaac had agreed to Alexander's leaving school at the age of only twelve, and becoming a musical apprentice to Heinrich Griesbach, the husband of Isaac's eldest daughter, Sophia. Heinrich came from a medical family of distinction, and Sophia had married above her station (the Herschels managed to pay for the requisite wedding entertainment only when the boys drew their army pay in advance) but Heinrich was the black sheep of his family and had been a humble fellow bandsman of Isaac's in the Hanoverian Guards. Somehow he had escaped internment, and through family influence he had been appointed town musician in a small town outside Hanover. There Heinrich was paid, believe it or not, in corn, and he supplemented this income by playing at functions in the surrounding villages and by making snuff.[17] He was also expected to take apprentices, and Alexander was the first of these.

Apprentices were traditionally treated as slave labor, but there seems to have been an element of vindictiveness in Heinrich's treatment of Alexander. Caroline was to see this as the reason for the difficult personality her brother displayed in adult life. Off duty, Alexander was very popular with the girls of the village and also with the landlord of the local inn, where one might find

> young men smoking tobacco and the daughters, maids and neighbouring young women assembled with their spinning wheels as busy spinning as possible, while the young men tried to keep pace with their pipes and some one or other telling stories.

Alexander's apprenticeship came to an end early in 1764 and he returned to Hanover, but the Herschels were living in cramped accommodation, and it was arranged for Alexander to board with the city musician. In return he "had little else to do but to give a dayly Lesson to an Apprentis and to blow an Coral from the Mrkt Turm," and he had just blown the chorale when William arrived home.

Most unfortunately for Caroline, William—her favorite brother, whom she had not seen since she was a child—had chosen to visit Hanover in the very Lent when she became fourteen years of age and so was due to be confirmed. What with the instruction in preparation for her Confirmation and First Communion, and the household chores that her heartless mother insisted she do, she saw very little of William. And her brother's visit ended on a melancholy note: it had become clear that William was committed to England, and Isaac realized that he was seeing his son for the last time.

Caroline was confirmed in the Garrison Church on Sunday, April 8, and her First Communion was scheduled for the following Sunday, the very day William was to depart. Caroline took her leave of her brother at 8 o'clock and set off for church in a black silk dress and carrying a bouquet of artificial flowers—the flowers that Sophia had carried at her wedding nine years earlier. At 11 o'clock, when the service was about to begin, the Postwagen carrying William away passed the open church door, and the postilion chose that moment to give "a smettering blast" on his horn. "Its effect on my chattered nerves," she recalled when she was in her nineties, "I will not attempt to describe."

A Turn for the Better

Back in Leeds, William resumed his musical duties. His records of this period are fragmentary, but the *Leedes Intelligencer* allows us to track his fluctuating fortunes. In the winter of 1764–65, for example, the number of those who took up a subscription to the concerts was so small that they had to be cancelled, for the income "would not have been sufficient for him to entertain the Company in so genteel a Manner as he could have wished to have done."[18]

Even when the concerts were in full swing, the industrious William made time for innumerable engagements elsewhere. Early in 1766 we find him spending two or three days a fortnight at Wheatley, the country seat of Sir Bryan Cook, who was an enthusiastic violinist and whose wife played the guitar. Some of Cook's relations would come over from Doncaster for morning concerts. All this was good for William but less good for his horse. "Having this time spent a whole week at Wheatly, my mare, standing idly in the stable, and being overfed by Sir Bryan's grooms died." He was at Wheatley on February 19, when his handwritten memoranda of the various places where he had engagements are interrupted by the entry "Wheatley. Observation of Venus." Five days later he is in Kirby: "Eclipse of the moon at 7 o'Clock A.M." But it would be years before astronomy featured again in his memoranda.

In March 1766 William moved his base to the town of Halifax, some sixteen miles from Leeds. There was to be a new organ built in the parish church, and in anticipation of the opening festivities, Joah Bates, the musically gifted son of the clerk of the parish church, was planning a performance of Handel's *Messiah*. The singers were to meet every second Friday for rehearsals, with William leading the orchestra; Joah Bates played the chamber organ and his brother the cello. William had designs on the post of organist, and the support of the Bates family would be crucial to his success. He took every opportunity to practice on other organs, and during the summer holidays in July, he deputized for the organists in both Leeds and Halifax while they were away.

Construction of the organ was opposed in the courts by many locals, who considered it a "heathenish thing,"[19] but the dispute was settled at last, and on August 19 the *Leedes Intelligencer* was able to announce: "An ORGANIST is wanted. Any Person who is inclined to offer himself as Candidate, may apply for further Particulars, to the rev. Mr. Bates at Halifax."

On August 27 the final rehearsal of *Messiah* took place. The oratorio was performed on the twenty-eighth and again on the twenty-ninth, and the competition for the post of organist took place on the thirtieth. William's allies, the Bates family, were in the church to offer their support, but William left nothing to chance. The instrument had no pedals, and so he used the trick of placing lead weights on lower keys so as to augment the harmony. The novelist Robert Southey tells the story. The seven candidates drew lots to decide the order in which they would perform. William was to play third, after Mr. Wrainwright of Manchester,

> whose finger was so rapid that old Snetzler, the organ-builder, ran about the church, exclaiming: "Te Tevel, te Tevel! he run over te keys like one cat; he vill not give my piphes room for to shpeak!"

Meanwhile a friend of Herschel's was standing with him in the middle aisle.

> "What chance have you", said I, "to follow this man?" He replied, "I don't know; I am sure fingers will not do". On which he ascended the organ-loft and produced from the organ so uncommon a fulness,—such a volume of slow, solemn harmony, that I could by no means account for the effect.
>
>   After this short extempore effusion, he finished with the old hundredth psalm-tune, which he played much better than his opponent. "Aye, Aye!" cried old Snetzler, "tish is very goot, very goot indeed; I will luf tish man for he gives my piphes room for to shpeak."[20]

Out of a field of seven candidates, the *Intelligencer* could report that "Mr. Herschel was unanimously elected Organist of the said Church. That Gentleman's great merit was abundantly evident from the important part he undertook, and so well performed, in the Oratorio."[21] William had played his cards well.

He was no doubt in euphoric mood at the time of the audition, because the previous day a letter had arrived from a Mrs. Julia De Chair in Bath, inviting him to accept the nomination for the post of organist in a chapel under construction there. So fashionable had Bath become that during the winter season the parish churches were unable to cope with the influx of aristocratic visitors, and private-enterprise chapels were being built where—for a fee—worshippers could pray without having to rub shoulders with the lower orders. The Rev. Dr. John De Chair had joined a banker friend in constructing what was to be called the Octagon Chapel, in Milsom Street. The central octagon was to be enclosed within a rectangle, and in the corners of the rectangle were to be fireplaces for the benefit of invalids. Snetzler had contracted to build the organ, and De Chair would conduct the services.[22]

How the De Chairs had come to hear of William is a mystery, but the fact that it was Mrs. De Chair who wrote must imply that she had met him somewhere. At all events, William—after suitable negotiations, no doubt over his stipend—was delighted to accept the chance to establish himself in Bath. For musicians there, the busy winter season offered rich pickings, second only to those of the metropolis itself.[23] Declining offers of an increase of salary at Halifax, he played the organ there for thirteen weeks, pocketed the thirteen guineas, and then departed for pastures new.

William Reaches the Promised Land

He arrived in Bath on December 9, 1766, and took rooms with a family called Harper. Their daughter Elizabeth was an attractive girl who sang as she sewed, and the resourceful William invented an excuse to spend time in her company by offering to give her lessons. To no avail, for "on disclosing his passion, he received no encouragement."[24]

It was important for William to announce his arrival on the Bath musical scene, and this he did on New Year's Day by promoting a private-enterprise "benefit" concert. Not many turned out in the depths of winter to hear this unknown artist, but those who did must have been mightily impressed when William played his own compositions on three different instruments: violin, oboe, and harpsichord. A couple of days later he set off on horseback for Leeds and Halifax to wind up his affairs there, and during his absence he advertised in the *Bath Chronicle* offering lessons in a whole range of musical instruments as well as in singing.

Such was the response that he quickly outgrew the Harper home. He decided to take a house of his own in Beauford Square, but he would need help to run it. In Leeds he had rented rooms with the Bulman family. But Mr. Bulman's business had failed and he was now out of work. By a stroke of good luck, De Chair and his partner were looking for a clerk to manage the finances of their chapel and to make sure the building was clean and warm. William successfully proposed Bulman for the job, and so the

Bulmans moved to Bath, where they were reunited with William in an arrangement that was to last for seven years.

Three weeks after his benefit concert, William was invited to join the band that played in the Assembly Room, Pump Room, and Baths. This would provide him with a second regular salary, to which would be added his income from freelance performing and teaching; but William already felt confident enough to decline what he saw as the subordinate post of a rank-and-file musician. It was then explained to him that no less a person than Thomas Linley, Sr., played in the band and that if William on occasion had a more lucrative offer to play elsewhere, he might send a deputy in his place. At this he accepted.

The occasions could be quite splendid. One delighted participant

found the most brilliant Assembly my Eyes ever beheld. The Elegance of the room, illuminated with 480 wax Candles, the prismatic colours of the Lustres, the blaze of Jewels, and the inconceivable Harmony of near 40 Musicians, some of whom are the finest hands on Europe, added to the rich attire of about 800 Gentlemen and Ladies, was, altogether, a scene of which no person who never saw it can form any adequate Idea—It began at half past 6 and ended at 10.[25]

In February William, frustrated at being an organist with no organ, wrote to Snetzler urging him "to hasten the work." At long last, on June 29, installation began. Jacob had arrived from Hanover a few days before, but Bath was out of season, and he soon took himself off to a country estate where he was to perform and teach music until the autumn. The official opening of the Octagon Chapel took place on August 4, and by that time William was engaging singers and performers for the inauguration of the organ, which was to follow in October. William was a man comfortable with all sections of society, and for his choir he enlisted "young workmen, carpenters and joiners, who had no previous notion of singing, but who, under his stimulating tuition, were soon able to render the choruses of various oratorios with success." For the opening on October 18 the oratorio, needless to say, was Handel's *Messiah*. William directed the performance, whose proceeds were for "the relief of the industrious poor," and Jacob played the organ. Between the second and third parts William himself performed an organ concerto, and as if all this were not enough for one day, in the evening he mounted a benefit concert on his own account, no doubt employing the singers he had brought to Bath for *Messiah*. And the next day *Messiah* was performed in the chapel for a second time, again for the industrious poor.

Until the arrival of William on the scene, Linley and his talented family had dominated musical life in Bath. In season, the spa town offered unlimited opportunities for the professional musician, but it was a cockpit of rivalries and became still more so in 1771, when the New Assembly Rooms were opened in competition with the existing rooms. Each year the season lasted from the autumn until Easter, and the musicians had to earn enough money in this period to last them a twelvemonth. William would teach for as much as forty-six hours in a single week. He also took part in lucrative private performances: in the winter of 1775–76, for example, the Marchioness of Lothian organized twenty successive Saturday evening parties, in her own house and in those of friends, at which William and a group of his pupils would perform. In addition there were benefit concerts (although these might prove a financial disaster if there were rival musical offerings that same evening), oratorios in the run-up to Easter, and concerts with the band, and meanwhile the choir of the Octagon Chapel had to be kept supplied with music and trained to perform it. And the sizeable—if less fashionable—seaport of Bristol was only a few miles away, and a concert given in Bath might be repeated next day in Bristol. Life in the season was frantic, and it is no wonder that tempers became frayed.

The opening of the New Assembly Rooms eventually led William to provoke a squabble that was a true tempest in a teapot.[26] William might reasonably have hoped to be appointed director of music in the new rooms, but it was his rival Linley who was so honored, and—to rub salt in the wound—William was merely to be one of the extra musicians needed on a Wednesday. One evening in January 1772 William found that Linley had failed to ensure that he had a stand on which to rest his music, and so he had no option but to place his music on the floor. William was outraged at what he saw as a public insult to the dignity of one of the city's premier musicians. When this happened a second time he lost all sense of proportion, walked out, and took an advertisement in the *Bath Chronicle* berating Linley for the "ungenteel treatment." Linley then took an advertisement ironically accepting that William's lack of a desk "must violently agitate the tender Sensibility of his Frame." In reply William informed the *Chronicle*'s readers that the "sensibility of Mr. Linley's *Frame*" was evidently not "tender enough to perceive the real Offence there is in leaving *any* Gentleman of the Band two *successive* Night's without a Desk." Before long, Linley was characterizing William as a man of "mean and contemptible Disposition," and was glad to inform him of "how very sincerely he despised him," comments that William ascribed to "that bitterness of Temper which is the general Attendant on low Cunning and dark Envy, when they are drawn out of their lurking Place and exposed to Public View." Calm, and a much-needed sense of proportion, returned with the end of the season.

Alexander in Bath

By this time Alexander had come to Bath and was lodging with William. Alexander's musical career had progressed since William's visit to Hanover in 1764. In the winter of 1765–66, he became first oboist in the regimental band of Prince Charles, who was a pupil of Jacob Herschel's and a brother-in-law to King George III. Alexander was thus well placed to secure one of the coveted vacancies in the Hanoverian Court Orchestra, for "Prins Carl being at Hannover, it was known that all vacancies would be filled with men from his favorit Band." Unfortunately, when the vacancy occurred Alexander was one of two equally favored candidates, and each had to agree to work for half salary. Caroline later commented: "This my Father foresaw would involve us in great difficulty, for he had early discovered (and not been sparing in admonishing) that Alex. was no economist and addicted to expensive pleasures."

In 1767 Isaac at last succumbed to the ill health that had plagued him since his early years of campaigning with the Guards. Jacob, the new head of the family, and his widowed mother now had to face life without Isaac. "About Alexander we had no fear that he would by practising a strict economy, and attending some of his late Father's Scholars, and others, do well enough till by an increase in salary his situation would become more easy."

Jacob, drawn to the rich pickings available in Bath to musicians of talent, now decided to pay William an extended visit, and this left Alexander unsupervised. Caroline, as she tells us,

> was extremely discomposed at seeing Alex, associating with young men who led him into all manner of expensive pleasures, which involved him in debts for the hire of Horses, and Carioles &c. and I

was (though he knew my inability of helping him) made a partaker in his fears that these scrapes should come to the knowledge of our Mother.

It was at this stage that Alexander began to display hints of the exceptional mechanical talent that he shared with William.

> My Mother Span, I was at work on a set of ruffles of dresden work for my brother Jacob, and Alex oftens sat by us and amused us and himself with making all sorts of things in pasteboard, or contriving to make a 12 hours Kuku Clock go a Week.

Jacob returned in July 1769 well pleased with his time in Bath, and a year later he was off again, this time with Alexander, who had been given two years' leave of absence from the court orchestra. Alexander was to stay in Bath not two, but forty-six, years. He quickly became a member of the band at the Orchard Street Theatre and performed there for most of his long residence in the town. Like his three brothers, Alexander had the ability to play any instrument he laid his hands on: a violinist and oboist in Hanover, in Bath he was known first as a clarinetist, and then as a cellist whose solos Caroline declares were "divine." But he was notorious for his insistence on strict tempo. He

> was a true German; being a strict timist, but scouting the more delicate refinements of Italy; in consequence of which, whenever Tenducci sang at any concert in Bath, he and Herschel were always sparring, as whenever the former wished to lengthen a note, or vary a little from the strict time, when the expression of the sentiment seemed to require it, which he used to signify to the band with a motion of his hand, Herschel would always keep on without varying an iota but keeping rigidly to the time, saying there was no pause or adagio marked.[27]

Alexander's talent for contriving gadgets was truly remarkable. In the years to come he would make William a clock that kept excellent time and another specially devised to help Caroline in her astronomical observations; and he even taught himself to become a brass worker to a professional standard. He was to be the third—but unsung—member of the Herschel astronomical team.

Caroline in Bath

William had all but forgotten his squabble with Linley by the time he arrived back from Hanover on August 27, 1772, with Caroline in tow.[28] It was not until the afternoon of the following day that his sister awoke from her sleep of exhaustion and found herself in Bath. Early the next morning it was down to business. Even before breakfast was finished, the irrepressible William was giving her a lesson in English, and then one in arithmetic. Because she was a girl, Caroline had not been taught arithmetic in the Garrison School, but in Bath William was to give her a weekly sum of money to pay for the housekeeping expenses, and she would be expected to account for it—in English. But in arithmetic William was not the best of teachers, and "we began generally where we should have ended; he supposing that I knew all that went before." Caroline did her best to learn the multiplication tables, but never succeeded, and in later life she would always carry a written copy with her. Geometry too she found a challenge, and William would impose sanctions if she got things wrong: "He used, when making me, a grown woman, acquainted with [mathematical figures], to make me sometimes fall short at dinner if I did not guess the angle right of the piece of pudding I was helping myself to!"

Her next lesson was in singing. William accompanied her on the harpsichord, and he encouraged her in a technique for practicing that strikes us as odd but was widely accepted at the time: singing with a gag in the mouth. And then "by way of relaxation we talked of Astronomy." The writing was on the wall.

Anxious that Caroline should happily integrate herself into the home he shared with Alexander and the Bulmans, William had portrayed Mrs. Bulman as someone who would be "a well-informed and well-meaning Friend," and her daughter, a few years younger than Caroline, "an agreeable companion." The Bulmans paid one-third of the thirty guineas rent and occupied part of the ground floor of the house in New King Street. Caroline and Alexander slept in the attic rooms, and William had the middle floor, the front room of which was "furnished in the newest and most handsomest stile" and was spacious enough for rehearsals and the performance of chamber music. Caroline was to get on well enough with Mrs. Bulman, although she doubted whether she would ever need some of the more sophisticated recipes that Mrs. Bulman taught her, but Miss Bulman she found to be "little better than an idiot." Caroline had grown up in a household of brothers, and she related well to the men around her, but women often taxed her slender reserve of patience.

This was especially the case with William's servant, "a hot-headed old Welsh woman" named Betty. Betty had until now worked happily enough under Mrs. Bulman's supervision, but Caroline gradually took over as her English improved. Relations between Caroline and Betty soon became fraught. Back in Hanover Anna had insisted on the highest standards, and Caroline was shocked at the state of William's cutlery and tea service and was determined to put matters to rights. "[T]hose articles which I was to take in charge such as Tea-things, glasses &c &c were nearly all destreued, Eivory hand[l]es of Kniefs & Fork and their Blades eaten up by rust, Hilters [handles] of the Tea-Urne &c were found in the Ash-hole." But it was the first time in her life that Caroline had given orders, and she had no idea how to do this in an acceptable manner. There was also the problem of meals. When the season started, William was out all day, so Caroline would ask at breakfast what he wanted for dinner, and then do her best to convey these instructions to Betty as coming from William rather than herself. Without success: "they were received with so much ill will" that Betty gave notice and departed at Christmas.

Mrs. Bulman recommended an agency to Caroline, and she hired another servant, on one month's notice on either side. This girl proved no more satisfactory than Betty; and so began a procession of comings and goings, some involving pickpockets and prostitutes, until a friend advised Caroline to take up references before offering anyone a job.

Shopping was equally traumatic, for this too was a new experience for Caroline, and her English consisted of no more than a few words. After only six weeks in Bath she "was sent alone among Fisherwomen, Butchers basket women &c and brought home whatever in my fright I could pick up." But unknown to her, Alexander was quietly shadowing her steps, ready to intervene should she get into real difficulties. Family legend was to insist that on one occasion she brought home a live suckling pig under the impression it was dead.

The winter of 1772–73 proved a difficult time for Caroline. William was so preoccupied with musical engagements that he could hear Caroline sing only while he was eating breakfast. This meal took place at seven o'clock, or even earlier, "much too early for me; for I would rather have remained up all night than be obliged to rise at so early an hour." His meal eaten, William would rush out of the house, leaving Caroline with instructions for long hours of tedious singing practice. Then in February came the sad news that their eldest sibling, Sophia, had been widowed, leaving her with six children to support, the youngest only a few months old. William and Alexander paid off her debts, but at such a distance, there was little else they could do to help.

Alexander too was out of the house for most of the day during the season, but (as Caroline records ruefully)

> if at any time he found me alone, it did me no good, for he never was of a cheerful disposition but always looking on the dark side of every thing, and I was much disheartened by his declaring it to be impossible for my Brother to teach me anything which would answer any other purpose but that of making me miserable.

But the season would not last for ever. Come Easter, William would surely be free to keep his promise to train her to sing.

2

1773–1778
Vocations in Conflict

William's Obsession with Astronomy

Unfortunately, by the spring of 1773 William was becoming obsessed with astronomy.[1] For some months past he had owned a copy of Robert Smith's two-volume *Opticks*, with its detailed account of how to make telescopes and its summary of what might be seen with the completed instruments. When he returned home of an evening, exhausted from up to twelve hours of music making, he would retire to his room with some milk or sago and spend much of the night reading Smith; and at breakfast Alexander would be expected to listen to a lecture on astronomy. One morning, the patient cellist might learn that the nearest stars were at least 400,000 times further from us than the Sun (but just how far no one knew). Another time the lecture would be on the rare "new stars," of which Smith could give three examples, or on the mysterious "lucid spots" in Orion, Andromeda, and elsewhere.

The season ended on Easter Sunday, April 11. Barely a week later William bought himself a quadrant for measuring the angular distance between two astronomical bodies, and on May 10 he purchased a copy of James Ferguson's *Astronomy*. Caroline's singing was slipping down his list of priorities.

Ferguson was a remarkable man. He was born in 1710 and was first employed as a shepherd boy, after a schooling that had lasted a mere three months. Then for fifteen years he worked in his native Scotland in a variety of domestic jobs, many of which exploited his aptitude for mechanical devices. In 1743 he took the bold step of moving to London, where his ingenious astronomical contrivances made a big impression in Royal Society circles. Ferguson found himself able to make a comfortable living as a public lecturer in science, and he became the unofficial "popularizer in residence" at the Court of George III, who in 1761 awarded him an annual salary of fifty pounds.

Ferguson had published his *Astronomy Explained upon Sir Isaac Newton's Principles* in 1756. If it was from Smith that William learned his telescope making, it was Ferguson who shaped his view of astronomy. Ferguson, being self-taught, was unconventional; William, also self-taught and much influenced by Ferguson, would be equally unconventional. Among almost all the astronomers of the day, professionals and amateurs alike, astronomy was the study of the bodies of the solar system—the Sun, the planets with their satellites, and the comets—and the stars did little more than provide an unchanging and therefore uninteresting backdrop to the orbits of these bodies. In the mid-seventeenth century Descartes had persuaded the best minds that the Sun was merely our nearest star and that all the stars were in fact free to move in infinite space; but when Ferguson first published his book, so little of interest was known about the stars that he gave them not so much as a single chapter. It was only in the second edition, published in 1757, that Ferguson added a dozen pages on "the fixed Stars." Ferguson's ideas helped shape the thinking of the apprentice astronomer in Bath, and in consequence they were to have an impact on the future development of astronomy far beyond anything Ferguson could have imagined.

Ferguson was a convinced believer in the Principle of Plenitude: God was omnipotent, and surely He would display his omnipotence when creating his universe, rather than limiting Himself unnecessarily. And so, if He put intelligent beings on planet Earth, why would He not do the same with the other planets, and with their satellites too—and with the planets and satellites of all the stars throughout the entire universe? Thus, in an imaginative passage that has become reality in our own time, Ferguson describes how the sky would appear to a "Lunarian" on the Moon.

Smith's "lucid spots" are described by Ferguson as "little whitish spots in the Heavens," and these nebulae (as they were usually called) apparently had no stars in them. The Andromeda Nebula, he says, "is liable to several changes, and is sometimes invisible," in which case it could not be a distant star system of vast dimensions. In addition there are "Cloudy, or nebulous Stars," which "look like dim Stars to the naked eye; but through a telescope they appear broad illuminated parts of the Sky"; the most remarkable is in Orion's Sword. These are spaces "in which there seems to be a perpetual uninterrupted day among numberless Worlds, which no human art ever can discover."

These few remarks provided an agenda for William's first modest steps in observational astronomy. But he could do nothing without first equipping himself to see the heavens. And so, just two weeks after purchasing the copy of Ferguson, he "bought an object glass of 10 feet focal length."

William's First Telescopes

Telescopes of the type introduced into astronomy by Galileo and still in use among amateurs today are "refractors." At the top end of the tube is a curved piece of glass, the lens, or "object glass," and as the light passes through this glass it is bent (refracted). The glass that William bought had been shaped so that, after the rays had traveled 10 feet down the tube, they would converge to form a tiny image of the planet or star. The observer would then use an eyepiece—essentially, a microscope—to magnify the image and make it available for study. Different eyepieces would give different magnifications.

The snag was that rays of different colors bend through (slightly) different angles as they pass through the object glass, so that the images in the different colors do not exactly coincide at the bottom of the tube, and this results in a blurring. Astronomers in the late seventeenth century found they could reduce the blurring by increasing the distance from the object glass to the eyepiece (the focal length). The Dutchman Christiaan Huygens carried the lengths of these "aerial telescopes" to extremes, so much so that at times he had to dispense with the tube altogether to avoid the tremors resulting from the wind. With such a telescope it required skill, patience, and luck to see anything at all, as William soon discovered. Caroline tells the story:

> It soon appeared that my Brother was not contented with knowing what former observers had seen, for he began to contrive a telescope of 18 or 20 feet long (I believe after Huyghens description) I was to amuse myself with making the tube of pasteboard against the glasses arrived from London, for at that time no Optician had settled at Bath; but when all was finished, no one besides my Brother could get a glimpse of Jupiter or Saturn, for the great length of the tube could not be kept in a straigh[t] line.

Caroline was not exaggerating, for William admits to attempting a refractor 30 feet in length, but ruefully records "the great trouble occasioned by such long tubes, which I found it almost impossible to manage."[2] Happily, there was a solution to the problem, as Newton had shown. All colors bounce off a mirror at the same angle, and so if a telescope dispenses with lenses and uses mirrors instead—if it is a "reflector"—the problem will not arise. In a reflector as designed by Newton, the rays passed down the whole length of the tube to a parabolic mirror at the bottom, and they were then reflected back to converge near the top. There Newton placed a second, small, flat mirror at an angle of forty-five degrees, and this reflected the image sideways, toward the eyepiece set in the side of the tube. Rather surprisingly, therefore, in a Newtonian reflector the observer is positioned near the top of the tube, facing sideways to the direction in which the telescope is pointing.

A fortnight after he bought his first object glass, William notes: "the use of a small reflector paid for." The word "small" was rarely allowed to feature for long in William's vocabulary, and soon he was writing off to London to ask the cost of a larger mirror. This was in the days before the technique of silver-on-glass had been devised, and mirrors were made of an alloy, usually copper and tin, known as "speculum metal," "speculum" being the Latin word for "mirror." The price quoted seemed to him exorbitant, but luckily he heard of a Quaker living in Bath who had been amusing himself by polishing mirrors for reflectors but who had now tired of the hobby. "Having found him out, he offered to let me have all his tools and some half finished mirrors, as he did not intend to do any more work of that kind."[3] After chapel on Sunday, September 22, 1773, Caroline went straight home but William called by appointment on the Quaker. "When I bought his apparatus, it was agreed that he should also show me the manner in which he had proceeded with grinding and polishing his mirrors; and going to work with these tools, I found no difficulty to do in a few days all what he could show me, his knowledge indeed being very confined."

Late in October, using the Quaker's preferred alloy of 32 parts of copper to 13 of tin and 1 of regulus of antimony, William had some disks cast for a modest 2-foot reflector, and a fortnight later more disks, this time of 5½ feet focal length; "and as soon as they were ground and figured as well as I could do them, I proceeded to the work of polishing."

William was now launched on his career as one of the outstanding telescope makers of all time. His first reflectors were of modest size—precision instruments for the study of our near neighbors, such as the Moon, the planets, and the brighter stars. Later he was to build himself telescopes with vast mirrors to collect the light from faint objects far away in deep space, what we might term "cosmological artillery." And he was to manufacture reflectors large and small for sale throughout Europe. But this was for the future. Meantime, he assumed as always that Caroline and Alexander shared his enthusiasm and so were only too eager to be of help.

In the case of Alexander this was true: oblivious to the danger of injury to his musician's hands, Alexander brought a huge turning machine from Bristol and installed it in one of the bedrooms, where he ground glasses and eyepieces. But Caroline, responsible for keeping their home in good order and at the same time eager to pursue her career as a singer, was near despair. "I saw almost every room turned into a workshop." On the other hand, more than most human beings she needed to feel needed, and now was her opportunity. To grind and polish a mirror to the parabolic shape required by the laws of optics was a hugely time-consuming business, and for William this was an art as much as a craft. To "feel" how the mirror was taking shape, he needed to have the metal in his hands without interruption until the task was completed. He would meanwhile become both hungry and bored. Once, "by way of keeping him alife," Caroline had to feed him food like a baby, for sixteen hours on end. More usually she read to him: *Don Quixote*, the *Arabian Nights*, Laurence Sterne's *Tristram Shandy*, novels by Henry Fielding, and so forth. And when there was "a fire to be kept in, and a dish of Coffe necessary during a long nights watching; I undertook with pleasure what others might have thought a hardship." Whether we are to take literally William's claim in 1781, "I have polished several hundreds of specula,"[4] or his even less plausible claim in 1785, "I have several times polished thirty hours without stopping,"[5] his skill in the craft of making telescopic mirrors would one day have momentous consequences.

Caroline's Traumatic Visit to London

William, meanwhile, was more than willing to assist Caroline in her career as a singer, provided this did not use up too much of that most precious of his nonrenewable resources, time. His tough little sister would need to acquire style and polish if she were to succeed as a soloist in the Handel oratorios he regularly promoted, and so he paid for her to have lessons twice a week from a noted teacher of dancing who was, Caroline tells us, "to drill me for a Gentlewoman." And an opportunity occurred for Caroline to hear the best singers in the land when one of William's pupils, a rich widow named Mrs. Colnbrook, was to visit London on business for a couple of weeks in January 1774. The chance was too good to miss, and so William arranged for Caroline to accompany her.

The searing experience that followed would live in Caroline's memory to her dying day.[6] A friend assured her—mistakenly—that before long William and Mrs. Colnbrook would get married and urged Caroline to be sure to make a good impression on her future sister-in-law. Caroline had been brought up to count every penny and had never given a tip in her life, but in London she was expected to accompany Mrs. Colnbrook to the theater or opera five or six nights a week, paying the entrance charges and all the incidentals. It was at William's expense, of course—he had given Caroline twelve guineas for the purpose—but that was little consolation to his frugal sister. The two ladies went to the opera at the Pantheon and to drama at Drury Lane and Covent Garden, but Caroline always regretted that they missed the final performance by that legendary Shakespearean actor, David Garrick. Caroline particularly enjoyed hearing Giuseppe Millico sing to his own accompaniment on the harp; Millico was a castrato, but it is impossible to tell from the autobiographies of the prim Caroline whether she understood how it was that he was able to sing at so high a pitch.

In the mornings Mrs. Colnbrook for some reason insisted on going to auctions, but when Caroline was asked her opinion of some lot, she was wise enough to remain noncommittal. Except once. Mrs. Colnbrook wished to buy a pair of carriage horses, and Caroline fell in love with two they were offered, ones "with white foreheads and Noses." The ladies went for a trial ride, and with Caroline's encouragement, Mrs. Colnbrook bought the pair and had them sent to Bath, only to discover they were both blind.

Caroline's patience during her ordeal was equal, if only just, to a stay of two weeks, and so were her funds. Then disaster struck: a snowstorm that made the roads in southern England impassable. The two ladies were marooned in London for a third week, then a fourth, and a fifth, and finally a sixth. Caroline ran out of money. She began by borrowing two guineas from Mrs. Colnbrook. Then, as William had told her to do if she found herself in difficulties, she applied for a loan from the Hanoverian agent in London, who lived, appropriately enough, in German Street. But each time she called, he was "not at home," and the suspicious Caroline began to think she was being fobbed off as a poor risk.

When at last the roads reopened, William got word through to both Caroline and the agent, and his harassed sister found herself in funds once more. But on her way back from German Street to meet up with Mrs. Colnbrook for the journey to Bath, Caroline decided she would thank William by buying him some of the Parmesan cheese he so much enjoyed, and in the process got hopelessly lost. When she finally found her way to their rendezvous in New Bond Street, she found Mrs. Colnbrook and her retinue standing by the carriage with the door open and making little secret of their exasperation.

To make up time they abandoned all thought of stopping for refreshments or to change the horses at Slough and covered the fifty miles to Newbury, the halfway stage to Bath, as fast as they could. They were accompanied on horseback by Mrs. Colnbrook's manservant, who on reaching Newbury, not unreasonably fainted. This sent Mrs. Colnbrook into hysterics, which lasted the whole evening.

The hysterics having run their course, things began to look up. Caroline gave Mrs. Colnbrook's servants the appropriate tips, and with this distasteful transaction behind her, she repaid the two guineas she owed their mistress. She spent the coach ride next day in happy anticipation of the welcome that awaited her in Bath after an absence from her brothers of six weeks. "But I was cruelly disappointed; for on arriving some hours after dark I was received by a huge blier-eyed Woman (a new Servant)." Alexander was away, and William ill in bed being nursed by Mrs. Bulman. Caroline herself was below par, suffering "the ill Efects acquired by six weeks Fasionable and harassing Town life," but she quickly recovered when she got back to her accustomed routine.

William Opens His First Observing Book

A month or so after Caroline's return, William's 5½-foot reflector had progressed to the stage where he felt able to open his first observing book.

On March 1, 1774, like so many amateurs before and since, he began by examining Saturn and the Orion Nebula. It was an interesting time to study Saturn, for the ring was almost edge-on to Earth. This happens every fifteen years or so, and then moons of Saturn that are normally lost in the glare of the ring become visible. William was to observe Saturn a number of times in the months ahead but without noticing anything of special interest; fifteen years later, things would be very different.

Needless to say, his unfolding passion for astronomy did not pass unnoticed among his fellow musicians. When the actor-singer John Bernard called on William at his home for a lesson, he found the room "heaped up with globes, maps, telescopes, reflectors, &c., under which his piano was hid, and the violoncello, like a discarded favourite, skulked away in a corner." Arriving on one cold and cloudy evening, Bernard was surprised to find William had positioned his music stand away from the warmth of the fire and near a window. As the lesson progressed, Bernard was declaiming his song with his eyes fixed on his music sheet, when William shouted out, "Beautiful, beautiful!" Bernard's pleasure at this compliment vanished when he realized that the exclamation was addressed not to himself but to a planet that had appeared from among the clouds.[7]

The musicians of Bath subjected William and his pupils to endless leg-pulling. Bernard's friends, tongue firmly in cheek, might demand that he outline the respective merits of the planetary systems of Tycho and Copernicus; or they would invite him to expound Newton's theory of fluxions. At other times they would ask if he had calculated the return of the recent comet. As for William himself, he might go off into a reverie about some astronomical problem in the middle of a rehearsal, and his fellow performers would say, "He's in the clouds again, he's star-gazing!" He was "called by the charitably disposed an eccentric," but what the less charitably disposed called him unfortunately is not recorded. However, some of his pupils approved of the transformation and would arrive for a music lesson but insist that he teach them astronomy instead.

It was in the summer of 1774 that William contrived his first acquaintance with a professional astronomer. When Bath was out of season, he sometimes arranged concerts in stately homes, and it may well have been at one of these that he was introduced to Thomas Hornsby, Savilian Professor of Astronomy at Oxford and the founder of the Radcliffe Observatory (then under construction). They evidently did not discuss astronomy face-to-face, but the encounter gave William the excuse to write to Hornsby in mid-December and ask for advice. He wanted to understand more about what happens when moons of Jupiter are eclipsed by their parent planet. This was a topic dear to Hornsby's heart, as shown by his recent purchase from the firm of Dollond of "a machine for limiting the aperture of the Telescope in observing Jupiter's Satellites."[8] He went to great pains to explain to William the geometry of these eclipses, for Hornsby could recognize untutored talent and William's letter showed him to be "very fond of the science of Astronomy."

By this time the Bulmans had returned to Leeds, and William had taken the opportunity to move a little way out of town, near Walcot Parade. His new landlord was a builder, and next to William's home was the builder's yard, with men eager to earn pocket money by doing odd jobs. William was as willing as Alexander to risk his musical career by handling tools capable of inflicting injury, and late one Saturday night he nearly paid the price. He and his brother were on their way home from a concert, and William was congratulating himself that the next day was a Sunday and so, except for attendance at chapel, he would be free to work on his telescopes. It then occurred to him that some tools needed sharpening. Their landlord had a grindstone in a public yard nearby, but they dare not be seen there in the morning, breaking the Sabbath. They must go to the yard at once, under cover of darkness; and so they took a lantern and the tools and went there at midnight. "[B]ut they were hardly gone when my Brother W^m was brought fainting back by Alex. with the loss of a nail of one of his fingers."

The success of the 5½-foot encouraged William to make himself a 7-foot with mirrors a little over 6 inches in diameter. The wooden mountings he contrived for these reflectors could be made by any competent carpenter; the excellence (or otherwise) of the telescope lay in its range of eyepieces, and more especially in the quality of the best available mirror of the required size, for these mirrors could be interchanged at will. William soon mastered the manufacture of eyepieces, and he now concentrated his efforts on the shaping and polishing of the mirrors. He ground and polished "many different object mirrors, keeping always the best of them for use, and working on the rest at leisure." He was rewarded on May 1, 1776, with a view of "the phenomena of Saturn's ring and two belts in great perfection." At the same time he was at work on a 10-foot reflector with mirrors 9 inches in diameter. (The leading professional maker, James Short, who had died eight years before, had advertised reflectors of up to

12-foot focal length and 18 inches aperture.[9]) William had read in Ferguson that the Earth and the Moon have a reciprocal relationship ("Our Earth is a Moon to the Moon"[10]), and so on the twenty-eighth, he turned his new 10-foot with its 9-inch mirrors on the Moon, wondering whether he might glimpse any signs of Ferguson's Lunarians. There was no evidence of intelligent activity, but he believed he could detect "*growing substances.*" In particular he suspected Mare Humorum to be a forest. Yet he was cautious, "since I can hardly imagine that any growing Substance would be long enough to be visible from the Earth to the Moon."

On July 30, 1776, there was an eclipse of the Moon, which William observed with great attention. By this time he had completed a 20-foot reflector with 12-inch mirrors, but the tube was crudely slung from a pole (see figure 4), and it seems he preferred to observe the eclipse with the 10-foot, which had a stable mounting. This event triggered an undated reflection, running to more than four folio pages, at the end of which he concluded: "were I to chuse between the Moon & earth I should not hesitate a moment to fix upon the Moon for my Habitation."[11]

The Brief Flourishing of Caroline's Musical Career

But meanwhile William had a living to earn from music. In September 1776 the two men who had built the Octagon Chapel parted company. De Chair's former partner now took sole control and appointed a new minister and a new organist. William and Caroline make no reference to this in their respective autobiographies, so we do not know if William had resigned or was dismissed (their silence suggests the latter), but the change could not have come at a more convenient time. Linley had recently departed Bath for the richer pickings to be had in London, and William had taken his place as the director of the New Assembly Rooms band. But in his heart he was by now an astronomer first and a musician second, and as director he was not a success.

In the spring of 1777, Caroline found herself deeply involved in the traditional Lenten oratorios. She copied innumerable parts for both instrumentalists and singers, she trained sections of the chorus, and she was herself to sing some of the treble solos. On March 5 she sang as a principal for the first time, in a performance of *Judas Maccabaeus* at the New Assembly Rooms. Her dress had cost William ten guineas. She sang again as

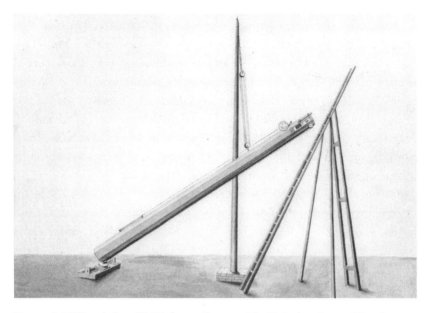

Figure 4. William's "small" 20-foot reflector with 12-inch mirrors. The observer was perched precariously in the dark at the top of the ladder, sometimes as much as 20 feet from the ground. From a drawing made in August 1783 by William Watson, RAS W.5/5, no. 4, courtesy of the Royal Astronomical Society.

principal a week later, and again a week after that, and the Marchioness of Lothian, no less, congratulated her on "speaking my words like an Englishwoman." A career beckoned. But as soon as the season ended, around Easter, the Herschel household reverted to astronomy.

Come the autumn, Linley was invited back to take charge of the New Assembly Rooms band. But at the turn of the year, Linley departed once more, and William found himself responsible for the concerts in both assembly rooms and in nearby Bristol as well.

April 15, 1778, was the day when Caroline reached the peak of her career as a singer. She performed in the New Assembly Rooms as first principal in *Messiah* (see figure 5), with such success that before she left the hall she was approached and offered an engagement in Birmingham. The few moments while she hesitated whether to accept or not were fateful: was she to set out on her own independent career as a singer, or would she be forever at the beck and call of William? But William it was who had earned her undying gratitude by rescuing her from slavery in Hanover, and she

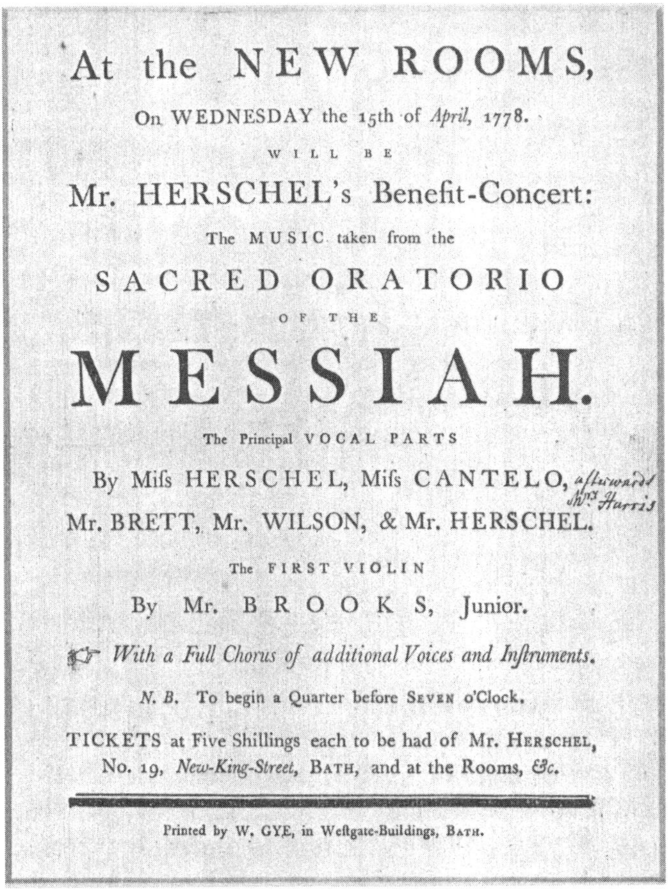

Figure 5. Poster advertising a performance of Handel's *Messiah* in Bath, April 15, 1778, with Caroline as first principal and William as fifth. It was after this performance that Caroline was offered, but declined, an invitation to perform in Birmingham, in a concert that could have launched her on a career as a singer of oratorios. Herschel Family Archives.

knew that he could never realize his ambitions—whatever they eventually proved to be—unless she was there to help. And so she declined: she would sing only when William was directing. It was her free choice, made before William had been so much as told of the offer, but being irrational like the rest of us, she would forever blame him for the ruin of her career. Her decision had trivial consequences for the history of music; the consequences for history of astronomy would be immense.

When the season ended, astronomy once again took over the Herschel household, and Caroline's career as a singer went into decline. To deflect the criticisms of his management and to meet the demand for singers of quality, William took to importing soloists from London, and before long, Caroline was doing no more than training and leading the trebles.

Dietrich's Flight from Hanover

Some months earlier, in July 1777, the Herschel household had been thrown into confusion by a letter from their mother in Hanover. Dietrich, youngest of her brood and now twenty-one years of age, had abruptly abandoned his enviable post as musician in the Hanoverian Court Orchestra and fled in secret with a lad of his own age, with the intention of taking ship for the East Indies. Could William or Alexander intercept him and bring him to his senses?

Bath being out of season, Alexander was away in nearby Bristol and out of immediate reach. William therefore immediately abandoned the lathe where he had been turning an eyepiece when Anna's letter arrived, and took the afternoon coach for London. There he was relieved to learn from the son of a Dutch merchant that no ship would be leaving for the East Indies until later in the year, so Dietrich must still be within reach. But where? William decided to continue to Hanover, where he learned that the runaways had got at least as far as Amsterdam. "I expect a Letter from Amsterdam every moment," William wrote to Caroline, "and we do already partly believe that he is already engaged to go to the Cape of Good Hope with a very musical gentleman; but nothing is certain yet." The family rallied round as they always did. Jacob covered for his brother's truancy from the court orchestra by securing him a year's leave of absence.

Matters then took a turn for the better. Dietrich's companion arrived back in Hanover. The runaways had gotten cold feet and abandoned their plans to go to distant parts; but rather than face his mother's wrath in Hanover, Dietrich had decided to continue to London, en route for Bath. William instructed the Hanoverian agent in London to give Dietrich whatever money he needed, and then he sent the good news to Caroline. Their mother was of course much relieved, and William took the opportunity to raise the matter of Caroline's continued stay in England. "Mama is extremely well and as I have represented things gives her consent to your

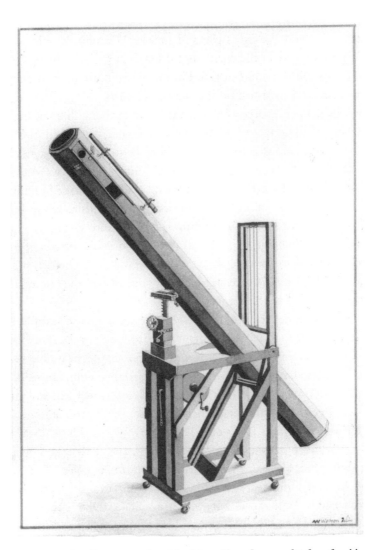

Figure 6. In 1776 William completed for himself a reflector of 7-foot focal length. The simple stand exemplifies William's good sense in these practical matters. The "most capital speculum" mirror he polished for this telescope in November 1778 would give him an advantage over all other observers, professional and amateur alike, and change the course of history. In later life William made many such 7-foot reflectors for sale, and an example is shown in plate 9: a carpenter would make the stand and William would concentrate on the optics—the eyepieces and mirrors on which the instrument's performance chiefly depended. From a drawing made in August 1783 by William Watson, RAS W.5/5, no. 3, courtesy of the Royal Astronomical Society.

staying in England as long as you and I please." Caroline should write to him in London care of the agent, and if Dietrich was still there, William would bring him to Bath. But Caroline had heard nothing from Dietrich; he had, it seemed, vanished off the face of the earth.

Eventually Caroline received word that he was lying ill at an inn near the Tower of London. With William away, she had to turn to Alexander, who fortunately was due to walk over from Bristol to visit her next day. In the morning she set out on foot to meet him with the news. Alexander swallowed a hasty breakfast, ordered horses, and set out for London. He arrived at Dietrich's bedside the following day, nursed his ailing brother for two weeks, and then brought him to Bath by easy stages. The doctor instructed Caroline that Dietrich was to have a diet of roasted apples and barley water; and as Caroline always did as she was told, Dietrich survived on nothing else until William arrived back, put a stop to the nonsense, and ordered the invalid to the family table.

It speaks volumes for the grimness of the regime of the family home back in Hanover that neither William nor Caroline nor Alexander seems to have uttered a single word of reproach to Dietrich for the immense trouble he had caused, nor did they suggest that he return home. Instead, William used his connections to get him musical engagements in and around Bath, and Dietrich was to spend there two happy and profitable years.

They were profitable years for William too, but in a different way. Dietrich was a lifelong enthusiast for entomology, and he taught William to collect butterflies. Indeed, after Dietrich's return to Hanover, William continued to send him specimens. Much more importantly, in the process Dietrich taught William how natural historians set about their business of gathering and classifying large numbers of specimens, and in the years to come William was to make himself the first natural historian of the heavens. This very phrase seemed a paradox at the time, for astronomy had been largely the mathematical study of the movements of a small number of familiar objects—Sun, Moon, Mercury, and so forth—each with its proper name and individual properties. By contrast, in the years to come William would collect and classify hundreds of double stars and thousands of nebulae, and he thereby transformed the very nature of the science.

William soon lost interest in the Lunarians, for evidence of their activities proved hard to come by; but he found a more appropriate challenge in measuring the heights of lunar mountains: by watching for when a mountain top first caught the rays of the Sun and measuring how far the mountain then was from the boundary between dark and light. Meanwhile he polished away at mirrors of all sizes, becoming ever more experienced at this difficult craft, and in November 1778 he achieved a mirror for his 7-foot reflector (figure 6) that he rightly considered "A most capital speculum": quite simply a masterpiece, the finest of its size—just over 6 inches—anywhere on earth. In studying the planets and the brighter stars, the excellence of this mirror was to give him a decisive advantage over all other astronomers, amateur and professional alike.

3

1779–1781
An Enthusiasm Shared

William Joins the Bath Philosophical Society

By this time William had made the acquaintance of the Astronomer Royal, Nevil Maskelyne. Maskelyne had been visiting a Bath friend, who thought his guest would be interested to meet the eccentric musician about whom everyone was talking. There followed what Caroline—who had not been introduced to their visitor—remembered as "several hours spiritted conversation," and she feared the men were quarrelling. But after William had seen Maskelyne to the door, he turned to Caroline with satisfaction and declared their future ally to be "a Deavil of a fellow." Other visitors to the Herschel home included Charles Blagden, later to be Secretary of the Royal Society; Alexander Aubert, doyen of English amateur observers; and Edward Pigott, another amateur soon to become famous for his work on variable stars.

But these were casual encounters. William's scientific isolation ended only in late December 1779, by which time he had relocated twice more and was now living in Rivers Street. One evening, when the Moon was to the front of his house, William took his telescope out into the street, set it up, and began observing. A passing coach stopped, and the occupant got out. He waited until William stood back from the eyepiece and then asked if he too might take a look. He offered his congratulations, and a conversation began that went on for hours.

The passerby was Dr. William Watson, a medical man like his more famous father of the same name. Both Watsons were Fellows of the Royal Society of London, and Watson Jr. had just attended the inaugural meeting of the Bath Philosophical Society, which was intended to provide a forum for local residents interested in science and related matters. It was the brainchild of Thomas Curtis, a governor of the Bath General Hospital, who on December 27 had suggested to another prominent citizen,

Edmund Rack, "the Establishment of a Select Literary Society for the purpose of discussing scientific and Phylosophical subjects and making experiments to illustrate them." The inaugural meeting had taken place the
very next day. Rack was elected secretary, membership was to be limited
to twenty-five, and the topics permitted for discussion were to be wide
ranging but should not include the contentious issues of "Law, Physic,
Divinity, and Politics."[1] The success of the Bath Philosophical Society, the
first properly constituted scientific society in the country outside London,
would depend upon its attracting members of the appropriate caliber. William looked to be a suitable candidate, and so Watson called on him next
day to introduce himself. He invited William to join the society and attend
the meeting on December 31.

If the society needed William, William needed the society. Its premises
were only a stone's throw from his home, and William at last found himself among a group of like-minded amateur enthusiasts for science. In the
society's books he had the nerve to list himself, not as a musician, believe
it or not, but as "optical instrument maker and mathematician."[2] The society met weekly in winter and fortnightly in summer, and in the coming
months William read the members no fewer than thirty-one papers on all
manner of topics, beginning—surprisingly enough—with corallines, an
interest for which Dietrich must have been responsible. Another paper he
read was on the heights of lunar mountains, a subject on which he could
speak with some modest authority, and a third reported his observations
of the famous variable star Mira Ceti, "the wonderful star in the Whale."
Watson thought both these papers contained enough of value for him to
forward them to the Royal Society in London, where they were read on
May 11, 1780. That day marked William's entrance onto the national
stage.

Both papers were then considered for publication in the society's *Philosophical Transactions*. The paper on the variable star raised no problem;
but the Committee on Papers, chaired by Maskelyne, decided—reasonably enough—that the one on lunar mountains should say more about the
instrument upon which the whole enterprise depended, the micrometer
that William had devised to measure angles on the lunar surface.

William was happy to oblige. But in his enthusiasm he also copied out
the pages concerning the inhabitants of the Moon that he had written in
the privacy of his observing book, including his stated preference for the

Moon as the place to live. And why not? Ferguson had devoted a lot of space in his book to the Lunarians, and this had proved no obstacle to his being elected a Fellow of the Royal Society and accepted at court as a respected astronomer.

Maskelyne's response left William in no doubt that this was not the way serious science was done—Lunarians were of no relevance whatsoever to the heights of lunar mountains. The offending passages were excised and the paper published. The Bath musician was evidently an astronomer to be reckoned with, but William had acquired detractors as well as admirers. In the years to come he was to become increasingly controversial, some seeing him as the greatest talent to appear on the astronomical scene in genera-tions, others condemning him as fit for the London mental asylum known as Bethlehem, or "Bedlam." Perhaps both were right.

William Begins His Search for Double Stars

A few weeks before his meeting with Watson, William had decided to devote what time he could spare from music and telescope making to fa-miliarizing himself with the starry sky. In what he called his first review, he had already made the acquaintance of the very brightest stars, but this had been a trivial exercise. Now in his second review he would observe in turn every one of the naked-eye stars (and beyond), and he would examine each of them to see whether it was indeed a single star, or two stars so close together as to appear as one—a "double star."

His motive was more than mere curiosity. One of the great unknowns in stellar astronomy was the distances that separate the solar system from its nearest neighbors among the stars. How to measure the distances was obvious. When you yourself are moving, the stationary objects around you appear to move, and the closer they are, the more they appear to move: their apparent movements provide the clue to how near they are. The Earth is in orbit around the Sun, moving from one side to the other every six months. As a result, we can expect the stars to appear to move back and forth in an annual cycle, the nearest moving the most; and since we know the diameter of the Earth's orbit, we can use simple trigonometry to convert the angle through which a star appears to move into the actual distance of the star.

Unfortunately, easier said than done. The nearest stars are in fact so far away that the movement we wish to measure is no more than the width of a coin at a distance of several miles. Worse still, this almost imperceptible movement takes months to go from one extreme to the other, during which time, changes in temperature and humidity may well cause warping in our measuring instrument and so invalidate our efforts. Further problems will result from changes in atmospheric refraction, and from all sorts of other complications too depressing to even mention.

Galileo had popularized an ingenious solution to the difficulty.[3] Suppose two stars lie in the same direction from us (so forming a double star), and suppose that one of the two—no doubt the fainter—is very much further away than the other. Warping in our instrument, changes in atmospheric refraction, and so forth, will affect the stars *equally*. By contrast, the Earth's orbit around the Sun will cause the stars to "move" by *different* amounts, each in proportion to its distance; and if the further star is very distant indeed, then for all practical purposes we can use it as a fixed reference point provided by a helpful Nature, against which to measure the apparent movement of the nearer star. The challenge will then reduce to monitoring the changes over the year in the tiny angle separating the two stars.

Knowing this, William set out to perform a service to astronomy by collecting specimens of double stars, to which other observers might then apply Galileo's method. But, unknown to him, there was a snag. Back in 1767 a paper had appeared in *Philosophical Transactions* written by the professor of geology at Cambridge, John Michell.[4] It occurred to Michell that there were surprisingly many double and multiple stars in the sky. Galileo's method depended upon the two stars of the double being at very different distances, one near and one far, in which case they appear to lie in the same direction from us purely by chance. Michell argued convincingly that there were simply too many of these alleged chance alignments of double and multiple stars for this to be the true explanation. Take the Pleiades cluster, for example. Can one reasonably suppose that the numerous stars of the cluster are at very different distances from us, and that it is purely by chance that they all lie in almost the same direction? Surely not.

William knew about Galileo's suggestion of using double stars to measure stellar distances, but fortunately he was unaware of the spanner that Michell had thrown into the works. With his usual dedication and commitment, on August 17, 1779, William set about examining each of the naked-eye stars, and more beside, to see whether the star was double. He began with Ursa Major and found that of the seven stars he studied, one was double. Cassiopeia came next (seven stars, of which two were double), then Draco (six, all single), and then the northern constellation of Ursa Minor. Its stars Beta, Gamma, Delta, and Epsilon were all single, but Alpha—otherwise Polaris, the Pole Star—was double. When, at the end of 1781, William had a catalogue of no fewer than 269 double stars to send to the Royal Society, it was Polaris that became something of a test case for William's credibility, for to all other observers this famous star appeared to be single. Surely it could not be that an eccentric Bath musician with a homemade reflector could see things hidden from the most experienced astronomers, armed with the best professionally made telescopes of the day.

William appealed to Alexander Aubert for support, for Aubert had an observatory lavishly equipped with instruments by the best makers of the period—Bird, Dollond, Ramsden, Short. He sent him maps with the newly discovered double stars marked in red and begged him

> to lend your assistance, that such facts as I have pointed out may not be discredited merely because they are uncommon. It would be hard to be condemned because I have tried to improve telescopes & practised continually to see with them. These instruments have play'd me so many tricks that I have at last found them out in many of their humours and have made them confess to me what they would have concealed, if I had not with such perseverance and patience courted them. I have tortured them with powers, flattered them with attendance to find out the critical moments when they would act, tried them with Specula of a short and of a long focus, a large aperture and a narrow one; it would be hard if they had not proved kind to me at last.[5]

He sent similar maps to Nevil Maskelyne at Greenwich and a list of some of his doubles to Thomas Hornsby at Oxford. A few weeks later, in March 1782, Aubert at last succeeded in confirming William's claim that the Pole Star is double, and the President of the Royal Society, Sir Joseph Banks, wrote personally to William to offer his congratulations.

Banks, Watson, and Aubert had all been resolved for some time to find a way to free William from the need to earn his living from music, so that he could devote himself to astronomy. They were committed allies, willing to go to any amount of trouble on his behalf. Indeed, Aubert was prepared to go even further—when critics mocked William's claim to have made eyepieces that magnified over six thousand times, he wrote to him:

> Go on my dear Sir with courage, mind not a few barking, jealous little puppies, a little time will clear up the matter & if it lays in my power you shall not be sent to Bedlam alone, for I incline much to be of the party.[6]

William's catalogue of 269 double stars, of which 227 proved to be new discoveries, demonstrated his credentials as a skilled and dedicated observer. As Maskelyne told him, it "must make every Astronomer tremble for his telescopes, lest they should not enable him to see what you describe." And by its very nature the catalogue set astronomy on a new course, for it was a work of natural history. William had collected specimens and classified them (by the degree of separation of the two stars) and ordered them within each class by the date of discovery. But this second review of the brighter stars had already yielded a far more sensational discovery.

William Discovers Uranus

Back in March 1781 the Herschels had moved once more, this time to 19 New King Street, and on the evening of Tuesday the thirteenth, William's review took him to the stars near Zeta in the constellation Taurus. Caroline was not with him, because he had recently made an ill-judged investment in a millinery shop; this was located some way from the center of town and so had failed to attract a sufficient clientele. The stock was being sold off, and Caroline had remained behind to make sure that the Herschels got their due share of the proceeds. As a result she was absent when William made an observation that was to change both their lives.

Because the shape of the mirror he had completed in November 1778 for his homemade 7-foot reflector was—as the Astronomer Royal was later to remark—of "extraordinary excellence," he realized the moment that he looked at what he expected to be just another star, that there was something odd about it. A true star would be so far away as to appear—in theory, at least—as a point of light. This object seemed to have a visible size (and a slightly blurred outline). Perhaps it was some sort of mysterious nebulous star; but if not, its visible size must mean that it was nearby, a member of the solar system. It was, he decided, probably a comet, for what else could it be? The decisive test would be whether it moved, for true stars are so distant that they appear to us to be virtually motionless. Four nights later he went back and looked at it again, and sure enough, it had moved. The next evening he invited Watson to join him at the telescope. Watson advised him to send a formal account to Watson's father, who lived in London and who was influential in Royal Society circles, and Watson meanwhile asked his father to "be so kind as to take proper measures to have this Paper read on Thursday so that Astronomers may have immediate notice of this event."[7]

And so it was that on March 22, just nine days after William had first glimpsed the anomalous object, astronomers at the Royal Society learned that the body was "much larger in Diameter but less luminous" than any comet and that "its body seemed very well defined having neither beard nor tail."

Comets, then as now, were of lively public interest. Some still saw them as harbingers of doom. The more scientific knew that the return of Halley's Comet at the predicted date had been a triumphant vindication of Newton's theory of gravitational attraction, and they were eager to calculate possible returns of other comets. There were astronomers who speculated about the role of comets in the economy of Nature—was it perhaps their purpose to refuel the Sun? And so when Maskelyne and Hornsby learned of William's discovery, they set about examining the supposed comet for themselves.

They immediately ran into difficulties. William was new to astronomy and did not yet know how to define the position of an object in the sky; and whereas his homemade telescope immediately showed him that the "comet" was anomalous when compared with the nearby stars, in Maskelyne and Hornsby's professionally made instruments, all the objects currently in Taurus looked physically much the same. Hornsby spent a frustrating week searching in vain for the supposed comet and was finally reduced to writing to William to ask for a sketch of its location. Maskelyne studied Taurus night after night, until at last, on April 3, he found an object that moved. These tiresome procedures can have done nothing to improve their tempers.

When at last they identified the object, they realized that it was indeed unlike any comet they had ever seen. Hornsby continued to regard it as a true comet, for the planets had been known since the dawn of history. But Maskelyne was not so sure; he immediately characterized it as "a comet or new planet," for if it was indeed a comet, it "seems a Comet of a new species, very like a fixt star." By April 23 his opinion had hardened: "It is as likely to be a regular planet moving in an orbit nearly circular round the sun as a Comet moving in a very excentric ellipsis."[8]

For a long time there had been talk of undiscovered planets. There was a good reason for this. In the solar system, the first four planets—Mercury, Venus, Earth, Mars—were closely packed; but then there was a huge gap between Mars and Jupiter, and astronomers had wondered whether the gap was in fact occupied by a planet as yet unseen. William's own early observations of the movement of the comet proved to be misleading (he mistakenly concluded that the object was increasing in apparent size and was therefore approaching Earth), and this did his reputation no good. When in the early summer of 1781 mathematicians were at last able to lay their hands on reliable observations, they found that the object was in fact a planet, but—astonishingly—one orbiting the Sun far out beyond Saturn: William was the first person in recorded history to discover a planet, and he had doubled the scale of the solar system. His allies in London would soon alert the king to the magnitude of his achievement.

William Attempts to Build a Monster Telescope

But William's ambitions lay beyond our little solar system, for he aimed at nothing less than "a knowledge of the construction of the heavens." His appetite for exploring the large-scale structure of the universe had been whetted in his first week of observing, when he had suspected the Orion Nebula of changing shape—if so, it could not be a vast star system, as some thought. He had examined this famous nebula a number of times since, but he needed reflectors with bigger mirrors—greater "light gathering power"—if he was to make a systematic study of such faint objects. So far his biggest mirror was 12 inches in diameter. A mirror 4 feet in diameter would be more to his liking—four times greater in diameter, and so sixteen times greater in surface area and "light gathering power." If he succeeded in making one, it would be the basic component of the biggest telescope in the world, and it would give its privileged owner—William the violinist—unique access to the remote regions of the universe.

His house in New King Street had a long narrow garden that extended south. He had no space in which to deploy a monster telescope toward the

east or west; but to the south, which is the crucial direction for astronomers in the northern hemisphere, he had an almost unobstructed view,
from only ten degrees above the horizon all the way up to the zenith.

He began to plan the reflector in January 1781.[9] As in all Newtonian
reflectors, the small secondary mirror would be positioned near the top
of the tube, to intercept the image reflected back up from the main mirror at the bottom and send it sideways to the eyepiece. The geometrical
relationship between the two mirrors was crucial, and the longer the tube,
the more likely that there would be distortion. He reckoned he would be
pushing his luck if he made the distance between them more than 30 feet,
and so he settled on this as the focal length. But even so, how could he
arrange to support a tube whose cross-section would be unprecedented in
size? The slim tube of his 20-foot was crudely slung by cords from the top
of a mast, and it was a nightmare to maneuver. Something much more
sophisticated would be needed.

What he devised was indeed sophisticated, but only in the sense that
the Rube Goldberg apparatuses of modern fiction are sophisticated. Confident that he would successfully cast the mirror, he went to all the trouble
and expense of building a complex framework to support the intended
tube (although he stopped short of making the tube itself). He set three
stout poles vertically in brick supports that were located on the ground
four feet from each other. These poles must have been well over thirty feet
in height. They were crowned by a circular metal cap. In the center of the
cap was a pivot, from which projected a horizontal arm that extended out-
ward beyond the poles, and at the end of the arm were pulleys and ropes to
support the top of the tube. Within the three vertical poles was an observing platform with safety railings. This could be raised or lowered to bring
the observer level with the eyepiece at the top of the tube.

At least, that was the intention. When the observer wished to examine
stars overhead, the 30-foot tube would be almost vertical, and then the
observing platform would need to be positioned an alarming 30 feet above
the ground. At all times the arm projecting from the cap would have to
support the weight of a tube 30 feet in length, which seems a big demand.
Then, whenever the observer decided to look at a different part of the sky,
he would need to come down to ground level and use the iron ring at the
back of the heavy mirror to lift it out of the tube. He was then somehow
to use the ropes and pulleys to alter the elevation of the tube, then pick

up the base and carry it to wherever was necessary for the tube to point in the new direction, and finally transport the mirror to its new location and drop it back in the tube.

The mounting (fortunately, we might think) was never put to the test, for the real challenge lay in the mirror. The first problem was its composition. William reckoned that a 4-foot mirror only one inch thick might weigh 575 pounds. Such a mirror would be fragile in the extreme, and it would have to withstand being dropped into place in the tube, yet to increase its thickness would push the price up to unacceptable levels. Perhaps some unconventional material, such as iron, would be less liable to accidental damage; and the gloss achieved by makers of candle snuffers and belt buckles suggested that an iron mirror might polish well enough.

Accordingly, early in February William bought "some very excellent iron in bars," and he took this to a local foundry and asked the founder to cast a disk for a 6-inch mirror as a trial of the metal. The furnace was heated up, but hours passed and still the iron refused to melt. Only after eight hours did it turn molten. And then, when the disk finally cooled, it turned out to be full of holes, useless as a telescopic mirror.

William blamed the founder, rather than his own choice of metal, and went to Bristol in search of a specialist founder accustomed to working with iron. This time the mirror looked more promising—it was still porous, but much less so—and William spent some hours in polishing it. He tried it out and found he could make out trees two miles away, but he had to admit that the mirror was not reflective enough to be of use in astronomy. He then tried working with a plate of wrought iron, and he persuaded himself that it showed signs—but no more—of taking a better polish. Again he blamed the founder for the lack of success.

These experiments were proving expensive and time consuming. "Nor indeed have I the patience to wait for uncertain results." On February 26, therefore, he reviewed the situation and came to two decisions. First, he would keep the mirror's focal length at 30 feet, but he would reduce the diameter from 4 to 3 feet, and the thickness from one to three-quarters of an inch. That would drastically reduce the weight (and therefore expense) of the metal he needed to 244¾ pounds. Second, he would experiment with the composition by making three small mirrors, each slightly different: respectively with 8, 9, and 10 parts of tin to 24 parts of copper.

These he would cast himself, and so the next day he began to prepare molds made of loam, which he would anneal with charcoal. Loam was an unlikely-sounding material for a mold, but the experiments he made were encouraging. However, the mirrors he cast were not a success. He tried other compositions, including the 1 part of tin to 4 of copper discussed by Pierre Joseph Macquer in his 1766 *Dictionnaire de Chymie*. This mirror turned out to be of excellent white texture, but it was extremely brittle.

His thoughts now turned to the great mirror itself. The disk would have to be professionally cast to a composition chosen by him; he himself would then grind it so as to hollow out the parabolic shape he needed, after which he would give the mirror the necessary polish. But then he hit a snag: it turned out that there was not a single foundry in the Bath region able to cast a 3-foot disk.

Undaunted, he decided he would cast the great mirror himself. What the other residents in New King Street thought about the prospect of the entire street going up in flames is not recorded, but the mad fiddler in no. 19 now "had a furnace and melting oven built in a proper room on the ground of my house." During its construction, "if a minute could but be spared in going from one Scholar to another, or giving one the slip, he called at home to see how the men went on." The room had what today's health-and-safety authorities would describe as two emergency exits, one into the house and the other into the garden, and because of this William and Alexander would live to cast another day.

William's cottage industry of mirror casting was as much a craft as a technology. When the mould had been hardened, "it was done over with ivory black and wetted with skimmed milk, as well as dried again before this last coat was laid on." For his next trial mirror he used seven pounds of copper and three pounds of tin, to which he added small pieces of candle and soap. "The metal when cast was of a very fine colour but very brittle. In pouring it into the mould it bubbled much like boiling water. . . In cooling several of the small edges broke off with little explosions or snapping noises." Many adjectives might be used to describe life in the Herschel household, but "dull" is not among them. While all this was going on William had discovered a planet and several double stars and conducted a number of oratorios.

Endless experiments with metal of different compositions went on throughout the summer. Not only did the planned composition have to be just right, but the molten copper had to be thoroughly mixed with the molten tin, a seriously hazardous requirement. Meanwhile William had to determine the exact curvature of the parabola needed to produce an image at a distance of thirty feet. This was a problem in geometry, but the geometry then had to be realized in a pewter pattern that would be used to give the mold an approximate shape, and so reduce the amount of grinding the disk would require.

William also had to devise how to insert a ring into the back of the disk, for the finished mirror would weigh in the region of a quarter of a ton and it would have to be regularly lifted out of the tube. And finally there was the problem of how to give the mirror its polish. Until now he had made only small mirrors, and these he had held in his hands while he polished; but for the heavy 3-foot mirror he would need to use a fixed polisher positioned on the ground. He was currently making telescopes for a couple of friends, and although their mirrors were small, polishing them gave him the opportunity to experiment.

William's recipe for the mold, inherited from the bell founders of centuries past, called for vast quantities of horse dung. Caroline and Alexander were recruited to take their turn in the pounding and sieving, and when William Watson, M.D., F.R.S., future knight of the realm, and future Mayor of Bath, was unwise enough to call while the work was in progress, he too was co-opted.

At long last, on August 11, 1781, William "cast the great metal." Over 500 pounds of molten materials ran quietly into the mold, but when it was nearly full, it cracked on one side, and the shape of the mirror was spoiled. But this hardly mattered, because the mirror itself later cracked on cooling. William thought this was because of the faulty composition of the metal, but Watson explained to him that a mirror needed to cool very gradually, for otherwise stresses would arise between the cooler exterior and the hot interior, with disastrous consequences.

Nothing daunted, William prepared to make a second attempt.

When everything was in readiness we put our 537,9 pounds of metal into the melting oven and gradually heated it; before it was sufficiently fluid for casting we perceived that some small quantity began to drop through the bottom of the furnace into the fire. The crack soon increased and the metal came out so fast that it ran out of the ash hole which was not lower than the stone floor of the room, when it came upon the pavement the flags began to crack and some of them to blow up, so that we found it necessary to keep at a proper distance and suffer the metal to take its own course.

Caroline gives a hint of the near-fatal drama: "both my Brothers, and the Caster and his men were obliged to run out at opposite doors." At this, even William admitted (temporary) defeat. The house is now the home of the Herschel Museum of Astronomy, and the scarred flagstones can be seen there to this day.

4

1781–1782
Royal Patronage

The Planet as Bait for a King

Not surprisingly, William's discovery of a planet made Herthel, Hertschel, Hertsthel, Hermstel, Herrschell, or perhaps even Mersthel[1] (no one seemed sure how to spell the name of this new arrival on the astronomical scene) famous throughout Europe. On May 2, 1781, he was an honored guest of Maskelyne's at Greenwich, on an evening when the viewing was particularly good; and in November 1781 the Royal Society awarded him their prestigious Copley Medal (Bath was in season, and so he traveled by the night coach to London to receive the award at 11 a.m. the following morning). A few days later he was elected a Fellow, "being well versed in Mathematics, Mechanics, and astronomy."[2] By unanimous vote he was exempted from paying the substantial annual contribution of thirty guineas, the stated reason being that the society knew that he would spend the money saved in this way in the pursuit of knowledge. This was tactfully put and no doubt true. But William was no wealthy gentleman, like the typical Fellow. He was a man who earned his living, if not with his hands then with his violinist's fingers, and it was his talent they wanted, not his money. As a further mark of esteem, he was allowed to postpone his actual admission until a day convenient to himself.

William's unprecedented discovery gave his allies in London the excuse they had been looking for. For centuries, well-established customs of courtly patronage had governed the relationship between an author or artist and his royal or aristocratic patron. The author or artist dedicated his work to the patron, and by permitting the dedication, the patron accepted the obligation to respond, usually with a gift of money or the conferring of a position at court. In astronomy, opportunities for patronage rarely occurred, but Galileo had seized the chance when he discovered the moons of Jupiter. He named them the Medicean Stars and dedicated the book announcing their discovery to Cosimo de' Medici, Grand Duke of Florence. The Grand Duke responded appropriately, by making Galileo his philosopher and mathematician.

By the late eighteenth century, the customs of patronage were becoming obsolete, but they had not yet vanished. During his first visit to William in Bath, Jacob had dedicated a set of sonatas to Queen Charlotte. He was summoned to court and performed so well there that his annual salary in the Hanoverian Court Orchestra was increased by 100 thalers, with the expectation of still better things to come.

Sir Joseph Banks, who as President of the Royal Society had influence at court and who was as shrewd an operator as you could meet, realized that William would put George III under a financial obligation if he called the planet "George." Colonel John Walsh of the Worcester Regiment was enlisted to point out to the king that William was a Hanoverian living in Britain, and that since George was both Elector of Hanover and King of Britain, he had a double motive for accepting the dedication. William's allies had no doubt that His Majesty would seize the bait with alacrity. Just imagine it! Forever after, the entire human race would know the planets of the solar system as Mercury, Venus, Earth, Mars, Jupiter, Saturn, George. What king could resist being in this divine company?

They were right, of course, but what might the king do in return? William was not a mathematician-astronomer like Maskelyne, dedicated to the problem of determining longitude at sea, nor a university professor like Hornsby, with an observatory equipped for precision observations, and in any case these men were in their prime and their positions were not in the gift of the Crown. However, Stephen Demainbray, the king's observer at Kew, was old, and his position was indeed at the disposal of the king. George, though something of a buffoon in popular imagination today, was in fact an intelligent and committed seeker after knowledge, and many years earlier he had learned that the planet Venus was to cross the face of the Sun in 1769. Expeditions were being sent worldwide to observe this very rare "transit," because mathematicians hoped to use the resulting observations to determine the distance scale of the solar system with unprecedented accuracy. George decided that he too would like to observe the transit, and so in 1768 he built himself an observatory in the grounds of Kew Palace, near Richmond to the west of London, and he appointed his old tutor Demainbray as observer.

In the English climate the king was tempting fate by building an observatory to observe an event lasting only minutes, but luck was on his side, and he and Demainbray were able to watch the tiny disk of Venus crossing the face of the Sun. But the observatory went into decline in the years that followed. Demainbray treated his post as a sinecure, doing little more than make the observations he needed to correct the official clocks in the neighborhood of Westminster. And now, in the first weeks of 1782, he was an old man. Banks prepared to coax the king into nominating William as Demainbray's successor.

But fate intervened. On the very morning when Banks planned to make his first approach to the king, news came that Demainbray had died. As Banks wrote to Watson on February 23, 1782:

> I wished the new star, so remarkable a phenomenon, to have been sacrificed somehow to the King. I thought how snug a place his Majesty's astronomer at Richmond is and have frequently talked to the King of Mr Herschel's extraordinary abilities. I knew Demainbray was old but as the Devil will have it he died last night. I was at the [King's] Levy this morning but did not receive any hopes. I fear [the time] has passed by which a well timed compliment might have helped if the old gentleman had chose to live long enough to have allowed us to have paid it.[3]

In fact King George had known about William and his discovery for many months, for the previous August he himself had spoken about it with Demainbray, who wrote to William: "His Majesty informed me last Friday, that you believed you had discovered a Comet, and that you intended to come to Richmond to ascertain its cituation with some of our fixed instruments." Demainbray assures William that he will be most welcome, but if he cannot accept the invitation, will he please inform Demainbray of the approximate position of "*your Comet or Planet.*"

William's portable reflector was not capable of measuring the position of the object in absolute terms. Kew, on the other hand, had precision-mounted instruments whose alignments were regularly checked against an obelisk mounted for the purpose due north of the observatory. But it would have been most uncharacteristic of William to wish to measure the position of the object himself, when the experienced Maskelyne could do this so much better at Greenwich. We can see the hand of Banks in all this. It was surely he who had informed the king of William's discovery, and he who had shrewdly invented a desire on the part of William to visit Kew, in the hope that this might bring William and the king face-to-face. It must have been Banks too who told the king about William's attempt to cast a 3-foot mirror, for the king evidently knew all about it when at length he met William.

For the king to appoint William to succeed Demainbray seemed an obvious step, yet George proved strangely reluctant when this was suggested to him. The reason, it turned out, was that he had promised the post to Demainbray's son. But it was only in July that he made this known, and until then Banks and Watson continued to hope the king would appoint William to Kew. Meanwhile George kept his own counsel, but because he had no idea what to do for William, he won himself some time by inviting him to bring his telescope to court.

Musical Distractions

Back in Bath, William was trying with minimal success to concentrate on his musical duties. Only four days after confirming that his "comet" belonged to the solar system, he directed an oratorio in Bath, and this was repeated in Bristol two days later. The performances were well received, but a year later, when rumors from London hinted at the exciting possibility of his being able to give up the musical grind and devote himself to his beloved astronomy, his mind was not on the job. On March 7, 1782, the local paper announced a forthcoming performance in Passion Week of Handel's *Jephtha*. The following week this was amended to *Samson*, and a week after that, only a few days before the performance itself, to *Judas Maccabaeus*. Signs of panic, one might think.

A performance of *Messiah* in Bristol shortly afterward was an unmitigated disaster. On the morning when William and Caroline were about to take a chaise to the rehearsal, carrying with them musical parts for nearly one hundred performers, William was deep in conversation with Watson about the impending summons to court. There then arrived one of William's nephews who played in Queen Charlotte's band at Windsor Castle, with "confirmation that his Oncle was expected with his instrument in town." Any prospect of William's focusing his mind on *Messiah* vanished. Caroline assembled the parts as best she could, and off they went to Bristol.

A letter to a Bristol newspaper describes the fiasco that followed. "Perhaps no audience was ever more impos'd on, or worse treated than that which Thursday night attended the performance of the *Messiah* at your theatre," it began.

> Many Gentlemen who went principally with a wish of hearing Mr. Tenducci, found themselves at the drawing up of the curtain (and not till then) disappointed—Hand-bills indeed were printed; but they were confin'd wholly to the company of the boxes—and so were only printed with a view to save the Manager's credit—It was expected and hop'd that some exertions would have been made by Mr. Rauzzini to compensate for Mr. Tenducci's absence—but that performer satisfied himself with singing one song, and joining, now and then, in a chorus.—Such was Mr. Herschel's eagerness to conclude the performance, that songs—duets—choruses, were omitted—the audience disgusted—and the band thrown into confusion. The first violin led off one air, while the violoncello had begun the accompaniment of another.
>
> The chorus singers were repeatedly at a loss whether to stand up or keep their seats; and Mr. Rauzzini had almost trampled Miss Storer to death, in endeavouring to sing from Mr. Croft's paper, instead of his own, which neither himself or the conductor of the band knew anything of.[4]

On May 1 a chastened William directed another performance of *Messiah*, this time to mark the inauguration of a new organ at St. James's, the largest parish church in Bath. Lessons had been learned from the Bristol debacle, and this time the choir was reinforced by choristers brought from Salisbury along with the famous singers of Lady Huntingdon's chapel. The caliber of the soloists meant that as usual there was no role for Caroline. But on Whitsunday, May 19, she sang the treble solo when one of William's anthems was performed in St. Margaret's Chapel in Bath with the composer at the organ. It was to prove the last occasion on which either of them performed in public.

"The Best Telescopes That Were Ever Made"

Early the following week William set off to meet Watson at the London home of Watson's father, where William was to stay. He had been told to bring his reflector for royal inspection, but the existing mounting of the 7-foot was too large to be taken by coach. The parts that mattered were the optics—mirrors and eyepieces—and so William devised a new stand and steps, which could be disassembled and carried in a box. He also took an atlas, his recently published catalogue of double stars, and an ambitious list of eight doubles he might show the king, including Alpha Herculis, which he considered "rather obscure and difficult." Seven of the doubles had been identified as such by earlier astronomers, and so did not call for William's unique skills as an observer, but he included the exceptionally handsome Gamma Leonis, whose double nature he had established only in February. He could not be sure that George was anything more than a typical amateur with little interest beyond glamorous objects such as Saturn's ring and the Orion Nebula, but it would be prudent to treat him as a serious observer, as indeed he was.[5]

On Friday William had dinner with a distinguished company that included Nevil Maskelyne, Alexander Aubert, and Colonel John Walsh. His first audience with the king took place on Saturday. William "met with a very gracious reception. I presented him with a drawing of the solar system," on which we can be sure William's planet was prominent. King George, still unclear as to how to resolve the Herschel problem, procrastinated once more. William was to take his reflector to Greenwich for appraisal by the Astronomer Royal and others, and then it was to go to the king's observatory at Kew for royal inspection.

William assembled his instrument at Greenwich on Wednesday May 29 and took the opportunity privately to assess the strength of the opposition. The previous spring he had observed with Maskelyne's telescopes during his visit there and thought the highly esteemed triple achromatic by Peter Dollond no better than his own "when the weather has not been favourable"! This time "I tryed the acchromatic telescope of Dʳ Maskelyne . . . with [magnification] 920 very strong aberration & ill defined. My reflector in tollerable fine weather is hardly so bad with 3168." Nor did William think much of Maskelyne's reflector of 6-foot focal length built by the great instrument maker James Short. The omens were good.

On Friday the king invited William to his private concert, and they chatted for half an hour. Then, on Saturday, William went to Greenwich, where Maskelyne awaited him with his assistant; the following night they were joined by Aubert.

> These two last nights I have been star gazing at Greenwich with Dr Maskelyne and Mr Aubert. We have compared our telescopes together, and mine was found very much superior to any of the Royal Observatory. Double stars they could not see with their instruments I had the pleasure to shew them very plainly, and my mechanism is so much approved of that Dr Maskelyne has already ordered a model to be taken from mine and a stand to be made by it to his reflector. He is however now so much out of love with his instrument that he begins to doubt whether it *deserves* a new stand.[6]

The following Tuesday William dined at Lord Palmerston's, and on Wednesday with Sir Joseph Banks. On Thursday he was at the king's concert. "As soon as the King saw me he came and spoke to me, about my telescope, but he has not yet fixed a time when he will see it." William left his telescope at Greenwich for the weekend and went to visit the private observatory owned by Aubert, who although an amateur was noted for the range and excellence of his telescopes. Now it was William's turn to make a trial:

> we have tried his Instruments upon the double stars and they would not at all perform what I had expected, so that I have no doubt but mine is better than any Mr Aubert has; and if that is the case I can now say that I absolutely have the best telescopes that were ever made.[7]

Which he had.

For the rest of June William languished in London, increasingly frustrated at the lack of progress. Bath was out of season, but he still had pupils, and Caroline was running out of excuses for his protracted absence. Whenever possible William went to Greenwich and made observations, and on Saturday June 15 an eminent group assembled there to look through his telescope, which was by now the talk of the scientific community. Maskelyne and Aubert were present, as were John Playfair, the Scottish mathematician and geologist; Anthony Shepherd, Plumian Professor of Astronomy at Cambridge; and John Arnold, the great watchmaker.

William was not alone in his frustration at the lack of tangible progress over his future. Watson in Bath was desperate for news, and he repeatedly urged William to take the initiative and broach the matter with the king. In particular, he should make it clear to the king that he would be honored to be appointed to Kew—for, Watson reminded him, protocol prevented the king from making any offer unless he knew in advance that it would be accepted.

> The King has shewn you every outward mark in his behaviour of predilection for you. But he might justly think that he ought previously to know that you are willing to accept of the place, before he makes you the offer. For want of knowing precisely your situation & wishes, how should he know but that you might be from your situation at Bath in such flourishing circumstances, as to make you above accepting of the Post of his Astronomer at Kew. . . . I should certainly take the first opportunity . . . humbly to request that you might succeed the late Dr Demainbray at Kew provided his Majesty thought of appointing [a] successor, & that you should look upon such a Post as the most happy event of your Life.[8]

Watson was also afraid that Hornsby, who had not been at Greenwich to see for himself the excellence of William's reflector, might have a poor opinion of William because of mistakes the inexperienced amateur had made in reporting of the positions of his planet. Hornsby, Watson insisted, must be converted into an ally before he had the opportunity to offer the king a damaging assessment; and if this meant delaying the Kew meeting with the king, so be it. In fact Hornsby, who was astonished at William's "extreme diligence" in the study of double stars, had no difficulty in recognizing a talent masked by limitations of education, and we hear no more of the supposed Hornsby problem.

Matters were delayed by a period of mourning at court—fortunately William owned the required clothes, which Caroline sent him from Bath—but the king was at his wits' end to know what offer to make. At one time the possibility of William's becoming royal astronomer at Hanover was mooted, but the proposed salary was no more than £100 per annum, less than a quarter of William's current earnings from music. But then an idea dawned on the royal consciousness.

The musical talent of the Herschels had passed down another generation to the five sons of William's eldest sister Sophia, two of whom were already in Queen Charlotte's band, which entertained guests during royal banquets (the other three boys were to join the band in the years to come). The problem for the king was what to do with guests once dinner was over. Now if he had an astronomer resident in the neighborhood of Windsor Castle, he could not only have his own private demonstrations whenever the mood took him, but he could send his dinner guests to look at the heavens through his astronomer's telescopes. It would not be lost on them that George was a most enlightened patron of science. William certainly had the required telescopes; but did he have the tact and courtly manners to act as host to visiting royalty and aristocrats? King George put him to the ultimate test: he invited William to bring his 7-foot reflector to Windsor Castle to show a selection of heavenly bodies to the royal family.

Hobnobbing with Royalty

The viewing sessions began on Tuesday July 2 and lasted at least three nights. William set up his homemade reflector alongside three of the king's instruments, all by leading makers. "my Telescope shewed the heavenly bodies much more distinct than the other Instruments. His Majesty saw [the double star] Epsilon Bootis with [magnification] 460 and the Pole Star with 932." It was only that March that Aubert, with his unrivalled arsenal of telescopes, had succeeded in verifying William's claim that the Pole Star was double; but with William's reflector, the king—inexperienced amateur observer though he was—was being invited to see this for himself. As William wrote to Caroline next day, "My Instrument gave a general satisfaction; the King has very good eyes and enjoys Observations with the Telescopes accordingly."

The next night William had an opportunity to demonstrate his skill in public relations.

This evening as the King & Queen are gone to Kew, the Princesses were desirous of seeing my Telescope, but wanted to know if it was possible to see without going out on the grass, and were much pleased when they heard that my telescope could be carried into any place they liked best to have it. About 8 o'clock it was moved into the Queen's Apartments and we waited some time in hopes of seeing Jupiter or Saturn. Mean while I shewed the Princesses & several other Ladies that were present, the Speculum, the Micrometers, the movements of the Telescope, and other things that seemed to excite their curiosity. When the evening appeared to be totally unpromising, I proposed an artificial Saturn as an object since we could not have the real one. I had beforehand prepared this little piece, as I guessed from the appearance of the weather in the afternoon [that] we should have no stars to look at. This being accepted with great pleasure, I had the lamps lighted up which illuminated the picture of a Saturn (cut out in pasteboard) at the bottom of the garden wall.

The effect was fine and so natural that the best astronomer might have been deceived. Their Royal Highnesses and other Ladies seemed to be much pleased with the artifice.[9]

The test passed, negotiations followed over terms and conditions. The king held a strong hand, for as Caroline tells us, her brother had no stomach for yet more music: "the prospect of entering again on the toils of teaching &c. which awaited him at home . . . appeared to him an intolerable waste of time." The deal they agreed was that "I should give up my musical profession, and, settling somewhere in the neighbourhood of Windsor, devote my time to astronomy." William was to be available to the royal family and their guests on request. In return he was to receive a "pension," or salary, of £200 a year. Watson alone of William's friends was told the exact amount, and as usual he was outraged on William's behalf. "Never bought Monarch honour so cheap!" But it was a fair sum. The salary of the Astronomer Royal was only £300, and *he* had to earn his money the hard way.

In due course William wrote to Banks a letter for *Philosophical Transactions* naming his planet Georgium Sidus. This echoed the Julium Sidus of one of Horace's Odes, and more importantly the name of Medicea Sidera, which Galileo gave to the moons of Jupiter when dedicating them to his patron. In Britain William was seen as having a free choice in the matter—Maskelyne invited him to "do the astronomical world the favor to

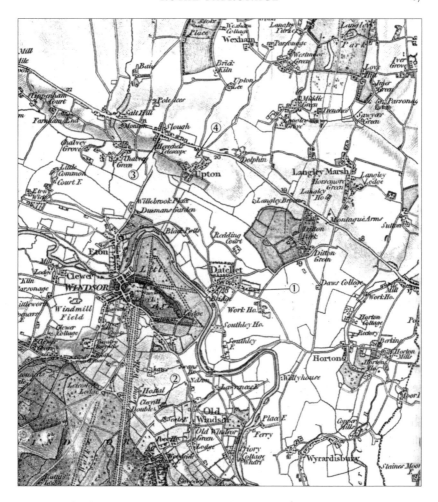

Figure 7. The first Ordnance Survey map of the Windsor area, published in 1830. (1) William and Caroline's home at Datchet, where they lived from August 1782 to June 1785; (2) Clay Hall, Old Windsor, which they occupied from June 1785 to March 1786; (3) "Herschels," Slough, where William lived from March 1786 until his death in August 1822; (4) Upton House, Upton, where Mary Pitt lived prior to her marriage to William in May 1788, and which she and William shared with "Herschels" as their joint residences for some time thereafter.

give a name to your new planet, which is entirely your own"—and there "Georgian Planet" was in common use until the middle of the nineteenth century. But his decision did not find favor among Continental astronomers; many of them at first called it Herschel, but eventually they adopted the proposal of the Berlin professor J. E. Bode that it be called Uranus—in mythology Uranus was the father of Saturn as Saturn was the father of Jupiter.

William scouted the Windsor area (figure 7) in search of a suitable home, and he found what he wanted in the village of Datchet, a couple of miles east of the castle. There is not the slightest hint that he consulted Caroline as to its suitability. Indeed, there is not the slightest hint that he asked her whether she was content to abandon her career in music and leave her numerous friends and acquaintances in Bath and transplant herself to the back of beyond, merely because this was what William wanted her to do. But this is what she did. William was a strange mixture of kindness and selfishness. When it came to money he was generosity itself. From Yorkshire he had sent money home to support his parents Isaac and Anna; he and Alexander were paying an annuity to Anna to compensate for the loss of Caroline's services; when Sophia was widowed he and Alexander had paid off her debts; when Jacob died suddenly he would decline his share in the estate in favor of Dietrich; and he would one day send money to Hanover to pay Alexander's funeral expenses. But he was ambitious, and when his ambitions were threatened, his thoughts were for himself.

He was equally neglectful in putting pen to paper, unless it was in his interest to do so. In his later life we find old friends and valued allies, such as William Watson, bemoaning the difficulty of keeping in touch with him; and when the king made him astronomer to the Court at Windsor, he could not even be bothered to write and tell his mother.

Anna was a country girl who had never learned to read or write. When Isaac was away with the Guards during the Seven Years War, Anna had kept in touch with him by dictating her letters to little Caroline. Now, when a grandson wrote from Windsor with news of William's appointment, her letter of congratulation was written in Dietrich's hand:

> Dear Son, I must at last really break through the long silence, for it seems as if you had entirely forgotten that pen & ink can be used for any other purpose than to write astronomical observations. . . . A letter from [her grandson] Griesbach informs us that the King has made you his Astronomer & that you are leaving Bath; it is most agreeable news for me as now you can live for the remainder of your days without care, although you will be no Croesus as regards riches.[10]

His mother was uneducated, but she was no fool.

5

1782–1783
"Astronomer to his Majesty"

"The Ruins of a Place"

For several days in late July a wagon stood outside the Herschel home in New King Street, Bath, while their possessions were loaded onto it. Not the least of their problems was the great pole that supported the 20-foot reflector. But by the evening of the twenty-ninth, the job was finished, and at two the following morning the wagon set off from Bath for Datchet. Nine hours later William boarded the London coach, which stopped for the night at one of the many coaching inns in the Speenhamland district of Newbury, halfway between Bath and London. The next afternoon the coach made its usual dinnertime stop at the tiny village of Slough, where the Great West Road was crossed by the route that headed north from nearby Windsor. William and the rest of the passengers alighted to take dinner at one of the coaching inns—perhaps the Crown on the southeast corner of the crossroads. William booked a room for the night, and when the coach resumed its journey it left without him.

The next morning he had some hours to kill. Slough was a tiny hamlet that took its name from the swampy ground that sometimes caused coaches to "slough," and it is more than likely that he went for a stroll south along the Windsor Road. If so, he would have passed a little cottage next to the inn, and two hundred yards further on a more substantial house then called the Grove. The cottage was occupied by an elderly widow—a wealthy widow, for she leased the inn and owned the cottage, the Grove, and all the properties in between (and much else beside). She would one day become William's mother-in-law, and the Grove would be his home for the last three and a half decades of his life.

Next day it was the turn of Caroline and Alexander to take the London coach. They too alighted at Slough, where William was waiting for them, and after dinner the three of them set off on foot south for Datchet, which they reached in less than an hour. The house in the village that William had recently rented stood empty, and so they took rooms in the Five Bells Inn (now the Royal Stag) next to the church. They awoke next morning to find that the wagon from Bath had safely arrived, and soon they were supervising the unloading.

In Bath Caroline had been accustomed to living in comfortable and well-appointed houses, and as she went round her new home, she was appalled.[1] It was "the ruins of a place which had once served as a hunting seat to some Gentleman and had not been inhabited for years," with rain coming in through the ceiling of every one of the four bedrooms. But even she had to concede that on the ground floor there was space enough: hall, two parlors, kitchen, larder, beer cellar, laundry, and washhouse.[2] The laundry, which opened out onto the garden, would make a good library. But what had attracted William had been the coach house and stables, which could be converted into workshops for the making of telescopes; and the spacious walled garden had room enough for him to erect his 20-foot reflector.

But there was not quite as much room as at first appeared. The garden was overgrown with weeds, and Alexander came within a whisker of falling down a concealed well. The brothers dared not venture further until men had been called in to mow the grass, "that they might see what ground they had to step upon."

On the recommendation of the king's upholsterer, William had hired a servant to help Caroline; but she was nowhere to be seen. It turned out that she was in prison for theft, and it was a fortnight before a trustworthy replacement could be found. Meanwhile, the gardener's elderly wife showed Caroline what few amenities Datchet had to offer. William had jokingly suggested that now they were in the country they could live off bacon and eggs, which must surely cost next to nothing. Quite the reverse: eggs cost two or three times as much as in Bath, and the butcher in Datchet overcharged and cheated on the weight. The only solution, Caroline decided, was to ride over to Windsor for her shopping. The problems did not stop there: dilapidated though the property was, it had some thirty windows, and so was highly rated for window tax. Their expenses had grown while their income had halved.

Alexander spent two months helping William get the garden into shape and the 20-foot telescope erected, but then it was time for him to return to

Bath for the beginning of the new season. In the past William and Caroline had twice extricated him from emotional entanglements, and they were full of foreboding as to how he would fare in Bath without them. Away in Hanover, Jacob and Dietrich were equally worried, and for exactly the same reason. Jacob tried to find Alexander an opening in the Hanoverian Court Orchestra, while William and Caroline urged him to move to London, where they would be within easy reach in case of problems. But to no avail: Alexander insisted on remaining in Bath; and within a year he married a widow called Margaret Smith—unhappily, if Caroline is to be believed. William got on with Margaret well enough when she and Alexander came to stay in the summer; but Caroline, as so often with female friends, found her a trial.

The silence in the house following Alexander's return to Bath brought home to Caroline the realities of her new situation. In Bath the Herschel household had resounded to music. Constant knocks on the front door had ushered in pupils for William to teach, or choristers for Caroline to train, while only minutes away had been a flourishing social life, its theatrical and musical entertainments surpassed only in London. Not any more: no more social life, no pupils, no choirs, no Assembly Rooms, and for Caroline no singing. Only an uncertain future, with William pursuing ill-defined duties as "Astronomer to his Majesty,"[3] and Caroline his housekeeper.

Ill-defined, but not negligible. The king was enjoying the novelty of having an astronomer at his beck and call, and William was regularly summoned to bring his 7-foot to Queen's Lodge at Windsor so that the royal family might view the heavens. He would then have to transport the reflector back to Datchet, so that he could spend what was left of the night enlarging his collection of double stars. It was expensive and it was time-consuming. Fortunately a break in the routine occurred in the second week of December when William returned to Bath for a few days, partly to collect the valuable metal fragments of the ill-fated 3-foot mirror. He and Alexander took the opportunity to cast another mirror for the 20-foot, which was just as well, for on New Year's Day the existing mirror cracked with the severity of the frost. Winters in those days could be very severe— on the night in question the thermometer registered eleven degrees Fahrenheit—and astronomical observing was a trial of endurance.

The king was by now very familiar with William's 7-foot reflector, but he had never looked through the 20-foot with its much greater "light gathering power," and for this he would himself have to make the journey to Datchet. And so, on December 1, 1782, began nearly four decades during which the Herschels would receive an unending stream of royal and aristocratic visitors come to see the stars—and to admire King George's patronage of astronomy. George's own fascination for the science of astronomy was entirely genuine (on one typical night the objects William showed him included the king's own planet, the celebrated double star Gamma in Virgo, and the star cluster we know as M35), and sometimes he would arrive at the Herschel home with only an equerry or two for company. We can picture William's maid interrupting her master with a message to the effect that the king was at the door, asking if he might look through William's telescopes. One Sunday it was the queen who proposed to visit, "provided you do not think it a sin to look at the planets on a Sunday's Evening." It was fortunate that William and Caroline would receive their sovereign or his consort only on nights fit for observing, for then there would be no rain coming through the ceilings of their home to dampen the royal pate.

William was at last a professional astronomer, but he was already in his mid-forties and had no time to lose. Within thirty-six hours of taking possession of the Datchet property, he resumed his search for double stars. For such a task—the scrutiny of stars that were bright and close at hand—his 7-foot with its precision mirror was the finest instrument in existence, and at the end of 1784 he would send to the Royal Society a second catalogue, this time of 434 doubles. But for the study of the faint objects that populated the remoter parts of the universe, the primary requirement was not precision but size: reflectors with huge mirrors that would collect enough light from these distant objects for them to register on the human eye. However, the attempt in Bath to cast a 3-foot mirror had ended in failure, the 20-foot with its 12-inch mirror was a promising if modest start (but its usefulness was limited by the primitive ladder on which the observer was perched at risk to life and limb), and a larger 20-foot with a stable mounting was still in the planning stage. For the time being, therefore, William persevered with heavenly bodies nearer to home.

The range and originality of the questions he asked himself in the coming months, in the time he could spare from double stars and the construction of the new 20-foot, is astonishing. In which direction are the Sun and its attendant planets moving through space? Do the light variations of the star Algol fall into a pattern? How do the spectra of the light of stars compare with that of the rainbow? And do the different colors travel through space at different speeds?

The Solar System's Journey through Space

Almost all of William's research publications in astronomy were to rely heavily on his own observations. But early in 1783 he sent to the Royal Society a paper that—uniquely—consisted of an analysis of data in the publications of other astronomers.[4]

That the stars are not "fixed" relative to each other on a heavenly sphere as the Greeks and medievals had believed, but are bodies free to move independently of each other in three-dimensional space, had been recognized for more than a century when William appeared on the scene. Motions in the line of sight—toward or away from the observer—would not be measurable until the development of astrophysics in the late nineteenth century, but in 1718 Edmond Halley had found that three individual stars were moving independently across the sky, each in a particular direction and at so many seconds of arc per century. Such a "proper motion" was identified and measured by comparing the present position of a star with its position at some time in the past; but the problem was that most past positions had been determined in ignorance of various complicating factors and so were of doubtful reliability. Nevertheless, as the eighteenth century wore on, an increasing number of plausible proper motions (in one if not both coordinates) were proposed, notably in 1781 by the French astronomer Jérôme de Lalande in a supplementary volume to the second edition of his *Astronomie*. The meticulous Nevil Maskelyne had set out a handful more in his 1776 volume of Greenwich observations, and these could certainly be relied on.

There were interesting implications in these proper motions. The Sun itself is a star, and if stars are moving relative to us, we are moving relative to them. In which direction then is the Sun (and the solar system) traveling through space?

How to determine this "solar apex" was well understood in principle. Consider how we explain the movements of the stars we see rotating overhead night by night. These have a common *pattern*; and we explain this pattern as caused by our movement rather than theirs (we conclude that we are viewing the stars from a rotating Earth). By analogy, William set himself to look for a pattern in the observed proper motions, and he would then explain it too as caused by our movement—this time, as the result of our viewing the stars in question from a solar system traveling through space.

He could understand the form the pattern would take by imagining he was on a walk that took him toward some trees. As he approached the trees, those to his left would appear to him to be moving further left, and those to his right, further right. William analyzed the data of proper motions listed by Maskelyne and Lalande accordingly and concluded that the solar system is moving in the direction of the star Lambda in the constellation of Hercules. This result appears little short of miraculous to our eyes, for Lambda Herculis is almost exactly the solar apex of modern astronomy. But when we examine his reasoning in detail, we find that by good luck he took one tiny positive datum to be reliable; whereas if it had been listed as negative rather than positive, as might very easily have happened, he would have arrived at a quite different apex. The great twentieth-century astrophysicist, Sir Arthur Eddington, was to remark of William, "It cannot be denied that he was given to jumping to conclusions in a way which, when it comes off, we describe as profound insight, and when it does not come off, we call wildcat speculation."[5] His identification of the solar apex was a bit of both.

William was never one to leave well enough alone. Two decades later, when many more data had become available, he set out to revise his earlier analysis—and to make some estimate of the actual speed with which the Sun and attendant planets are moving. In 1805 and 1806 he published his conclusions in *Philosophical Transactions*, in two papers totaling no fewer than fifty-seven pages. But the simple pattern he had previously discovered in the handful of proper motions of stars then known was now largely lost to sight, and although he managed to reason his way to a revised direction, his argument was convoluted. Worse still was his attempt to determine the speed with which the solar system is traveling. He argued that our speed through space is revealed to us by the apparent speeds of the stars we see around us, yet the only clue he had to a star's speed was obtained by multiplying the star's proper motion across the sky by its distance from us. But these distances were in fact unknown to him; he could only hope that "brightness implies nearness," that the stars are physically uniform and so their distances are revealed in how bright they look. Unfortunately the stars actually vary enormously, some being millions of times more luminous than others, and so the principle underlying William's analysis was hopelessly misconceived.

Algol and Other Stars That Fluctuate in Brightness

It is no surprise to find that changes in the positions of the stars went undetected until the eighteenth century, because they lie at very great distances and so their proper motions are barely perceptible; but the failure of early observers to notice changes in brightness is less easily explained. It is due partly to a joke played on us by Nature (for it happens that few of the naked-eye stars vary to any great extent) and partly to the tendency of us humans to see what we expect to see. But no one could fail to notice the (supernova) explosion of 1572, which remained visible for many months, and which Tycho Brahe proved was in the heavens rather than in the atmosphere of the Earth.[6]

Over time other temporary "novae" appeared, including one in the constellation Cetus seen in 1638 by the Frisian astronomer J. P. Holwarda. The pages of Holwarda's account of his discovery had already been printed when, to his astonishment, the star reappeared, something that had never been known before. By the 1660s it was established that Mira Ceti, "the wonderful star in the Whale," reached a maximum brightness every eleven months or so, although its actual light curve (as we would term it) varied from one cycle to the next.

Sunspots had by then been known for half a century. They rotated with the Sun, but sometimes there were more of them and sometimes fewer. Now if Mira was a star with very large "sunspots"—dark patches—and was rotating every eleven months, that would explain why its brightness rose and fell over this span of time; and if there were sometimes more and sometimes fewer of these "sunspots," that would explain the irregularities observed during these cycles.

Unfortunately this explanation was simply too flexible to be tested and so either confirmed or refuted—it could explain almost any changes in the brightness of any star. Worse still, if an observer announced that he had noticed that a particular star had varied in brightness, there was no obvious way for other observers to pass judgment on his claim, and such claims therefore became a facile way to make a reputation. Indeed, one observer in 1670 detailed a number of (alleged) changes in individual stars and then blithely added: "I have observed many more changes among the Fixed Stars, even to the Number of a Hundred." Not surprisingly, by the end of the seventeenth century the study of variable stars had fallen into disrepute.

Its revival was due mainly to Edward Pigott, who in 1778 had been one of the first amateur astronomers to call on William at his Bath home. Two years later, William's first paper appeared in *Philosophical Transactions*. Its subject was the variations in Mira Ceti, and it may be that it was this paper that triggered Pigott's interest in variable stars. At all events, that same year Pigott's father, Nathaniel, a skilled surveyor and an enthusiast for old-fashioned astronomy, moved his family to York, where he used his wealth to create a conventional observatory with precision instruments of high quality. Edward Pigott, possibly to annoy his father, decided instead to devote himself to the unfashionable study of variable stars, and late in 1781 he began to comb the seventeenth-century literature to see what he could find out about them.[7]

As chance would have it, a stone's throw from the Pigott home in York lived a seventeen-year-old deaf-mute, John Goodricke. Like so many teenagers Goodricke was fascinated by astronomy, Pigott was of a kindly disposition and in need of scientific companionship, and the two became firm friends. They could communicate only in writing, even when in the same room together, and on the day when they had "too warm a dispute" we can picture them exchanging penciled notes of ever-increasing bluntness.

Pigott taught Goodricke how to monitor possible fluctuations in any star that was suspected of being variable. They were to list the star and its neighbors in strict order of brightness, and later they would return to the region to see whether the ordering embodied in the list was still valid; if not, it would be proof that a change had occurred. Each would observe from his own home, and they would meet up next day to compare notes. Mira Ceti was of course near the top of their list of suspects, as was Algol, or Beta Persei, reported in both 1667 and 1670 as being much fainter than usual, of only fourth magnitude instead of second.

On November 7, 1782, Goodricke made a routine check on Algol and found it "same as before." But five nights later he could hardly believe his eyes when he found it reduced to fourth magnitude—so rapid a change was without precedent. The next day he shared his excitement with Pigott, and that evening they both examined the star. Both found it back to its usual brightness. Naturally, in the nights ahead Algol was high on their agendas, but it seemed always to be second magnitude. However, on December 28 both men were astonished to see Algol start out the evening at only fourth magnitude and slowly brighten to second magnitude before their very eyes. Nothing like this had ever been seen before.

In a flash of brilliance an explanation occurred to Pigott. Perhaps Algol was being orbited by a large planet and this sometimes passed between Earth and its parent star. In short, he suspected Algol was being eclipsed. If so, he and Goodricke must establish the number of days the supposed planet took to orbit Algol, so that they could predict when future eclipses might be expected to occur, and thereby confirm (or refute) the hypothesis. Goodricke had observed the eclipse (if such it was) forty-six days before, and so Pigott made calculations of future "eclipses" on the assumption that the planet took either forty-six days, or half that time, to orbit Algol.

They both kept close watch on Algol in the weeks to come, and as they observed more and more of the "eclipses," so the longest possible orbital period of the planet grew shorter and shorter, until by April it had decreased to less than three days. Nothing remotely like it had been known in astronomy before.

Pigott now decided to bring both the Astronomer Royal and William into the picture, and so he wrote to Maskelyne, asking him to pass the word to William. Maskelyne's letter reached Datchet on April 27, 1783, and William at once began to keep watch on the star. Meanwhile London scientific circles were abuzz with the news, and on the thirtieth, Sir Joseph Banks also wrote to William summarizing Pigott's hypothesis: "that one of the fixed stars has a planet revolving round it so exactly in the plane of our orbit as to eclipse the star partially each revolution." Later the same day Banks sent word to ask if he might come to Datchet that evening to view Algol, which he did; but without success, for evidently they had slightly miscalculated, and the drop in brightness took place just after they left off observing. On May 2 Banks wrote to William again, to say that he had learned that the discovery was due not to Pigott but to "a deaf & dumb man . . . who has for some years amused himself with astronomy." This was an extraordinary act of generosity on Pigott's part, for a surviving slip of paper in Pigott's hand proves that the hypothesis of an eclipsing planet was his rather than his handicapped friend's; and as a result York University today has a Goodricke College while Pigott is largely forgotten.

On May 8 William went up to London for the regular meeting of the Royal Society, and not surprisingly he armed himself with a summary of his observations of Algol. When he arrived it was already after four o'clock, and so he hurried to the Crown and Anchor, where the Fellows were accustomed to dine before a meeting. He was greeted with demands as to whether there was truth in the rumors of Algol's rapid changes in brightness. William had at no time been asked to keep the news to himself—indeed it appeared to be public knowledge—and so he handed his summary to the president, Banks, who read it to the assembled diners. There seemed no reason why it should not also be read at the formal meeting of the society that followed.

William later learned to his dismay that he had unintentionally stolen Goodricke's thunder. Pigott had left York some weeks before, leaving Goodricke to prepare a paper for *Philosophical Transactions*; but Goodricke, a teenager who had never published before, delayed until May 12 before putting the paper in the post. Even then he hedged his bets and offered two alternative explanations for the drop in brightness: eclipse by a planet (which we now know to be the correct explanation) or the century-old theory of dark patches. Fortunately Pigott was a long-standing friend of William's, and he and Goodricke accepted William's explanation of the misunderstanding. Curiously, the two York friends eventually abandoned the eclipse explanation in favor of dark patches, probably because they were deceived into thinking the reductions in brightness were not occurring with the required uniformity.

Goodricke and Pigott went on to discover three more short-period variables (none of them in fact due to eclipses), but their partnership came to an end early in 1786 when the Pigotts moved away from York. Shortly thereafter Goodricke died, supposedly from exposure to the night air. He was just twenty-one years of age and had been an F.R.S. for just two weeks.

The York astronomers had enriched astronomy with a new class of star: variables of very short period. But they had also provided a technique for the identification of variables of whatever period, for their sequences of stars arranged in strict order of brightness would force a variable to reveal itself by sometimes disturbing the sequence in which it occurred. In the closing years of the century, William put this method into effect as only he could. He undertook a mammoth program of observations that led to catalogues in which he left to posterity a comprehensive record of the brightnesses of stars as they were in his day. A century later the Harvard astronomer E. C. Pickering commented: "Herschel furnished observations of nearly 3000 stars, from which their magnitudes a hundred years ago can now be determined with an accuracy approaching that of the best modern catalogues."[8] It was a fitting memorial to Goodricke.

The Spectra of Starlight

Sunlight passing through a prism gave the familiar rainbow-like spectrum, and the stars were distant suns, so what would happen when starlight passed through a prism? Prismatic analysis of starlight would become the basis of astrophysics in the late nineteenth century. By then laboratory experiments had shown that by passing light through a prism one can discover the chemical composition of the source of the light. When applied to astronomy, this procedure invalidated the French philosopher Auguste Comte's 1835 dictum, that we shall never know what the stars are made of. But in William's day the spectrum of a star could be little more than a curiosity of nature, which no one could interpret.

Nevertheless, some of his friends pressed him to see what would happen, for William's telescopes could collect enough light from a bright star to yield a visible spectrum. One of these friends was Thomas Collinson, who stayed the night at Datchet in the spring of 1783. Collinson later wrote:

> The rising of Sirius reminded me of what I suggested to you at the time I had the pleasurable advantage of sleeping under your roof at Datchet—I mean that of refracting or separating the rays of a Star by the application of a Prism to your telescope.[9]

By then William had in fact already done as Collinson suggested. He had always been fascinated by the variety of colors to be seen in stars, and in particular by the purity of light in what he termed "the garnet star," Mu Cephei. He therefore pointed the 10-foot reflector he had made back in 1776 (the largest of his telescopes thus far to have a stable mounting) toward the constellation Cepheus. Holding a prism with his finger and thumb in between his eye and the eyepiece, he compared the spectrum of Mu with that of the brighter star Alpha Cephei. Alpha's spectrum included red, orange, yellow, green, blue, purple, and violet. The garnet star's was simpler: only red, yellow, and green, with perhaps a little orange, but no blue, purple, or violet. This was perhaps rather as one might expect, and there was little William could do with the information. But his observations that night, May 21, 1783, inaugurated the prehistory of astrophysics.

In April 1798, William's curiosity for stellar spectra would revive, and this time he contrived to attach to the eyepiece of his 7-foot reflector a device that held the prism while allowing it to rotate. This enabled him to make a careful examination of the spectra of six of the brightest stars. That of Arcturus, for example, he found to have proportionally more red and orange, but less yellow, than that of Sirius; Aldebaran's "contains much orange and very little yellow." It was the first such investigation, but he could do no more than record his findings.

Do Different Colors Travel at Different Speeds?

Newton had shown that sunlight is not simple but is compounded from the various colors that we see in the rainbow. When sunlight is passed through a prism, these colors are bent (refracted) through slightly different angles, red being bent the least and violet the most. But why was this? Newton thought this might happen because the particles of light were of different sizes, or densities. But in 1752 the Marquis de Courtivron suggested red light is the least refracted because it is traveling faster than the other colors and is therefore the most difficult for the prism to deflect from its original direction—which sounded plausible.[10] In a manuscript from William's Bath Society days, he notes that he intends "To examine the arguments for a difference of velocity in different coloured light."

One way to test this was by observing an eclipse of a star by a planet, rare though these are. At the instant when the star emerged from eclipse, all the different colors that made up the star's light would simultaneously begin their journey from the edge of the planet to the observer on Earth. If red light traveled the fastest, it would arrive first and the star would briefly appear to be red. Then, as the other colors also arrived, the star would revert to its normal color.

The greater the distance of the planet from Earth, the greater the time intervals between the arrival at Earth of the various colors, and the easier it would be to detect the initial redness. William's own planet was by far the furthest from Earth, and in November 1783 Patrick Wilson, professor of practical astronomy at Glasgow University, suggested to William that astronomers should watch for occasions when his planet passed in front of a star.

In his reply, William was discouraging. He believed that God had made the planet fit for habitation by intelligent beings, and so he was strongly inclined to think it must have an atmosphere. Such an atmosphere would blur the reddening effect. It chanced that a couple of months earlier Samuel Vince of Cambridge had written to him discussing other, similar tests, including observations of eclipses of the moons of Jupiter, which are happening all the time. William had discouraged Vince, and for the same reason, namely that the reddening would be blurred by the atmosphere "probably surrounding every celestial body" that was capable of causing an eclipse. But, he told Vince, despite his reservations he had made a few observations of eclipses of moons of Jupiter with just this test in mind.

He had made one such observation, of the eclipse of Jupiter's third moon, on October 28, 1783. And what do we read when we ferret out William's own note of what he saw? At 5:26 a.m. the moon "was redish"; at 5:28 "it had nearly regained all its light." His observation precisely confirmed the prediction of the theory that red light travels fastest. But his belief in a divinely supplied atmosphere for the inhabitants of Jupiter—an atmosphere that would complicate and even invalidate the light test—closed his mind to the implications of what he had seen.

6

1783–1785
The Construction of the Heavens

Caroline's First Nebulae

On arrival at Datchet, William had faced a number of problems, and one of these was how to keep Caroline usefully occupied. Her career as a solo singer had been abruptly terminated, and she no longer had choirs to train; in fact she had nothing to do beyond managing their little household. Was it possible that she could be coaxed into sharing her brother's passion for astronomy? William had tried to teach her the constellations on their journey across Holland on their way from Hanover, and at Bath he had made her a modest reflector, although there is no record of her ever having used it. Now he rigged up a little refractor to pivot around a vertical spindle and told her to go out and find interesting things with it: double stars, comets, nebulae, anything unusual (figure 8). She was to set the tube at a given elevation and then rotate it around the spindle, thereby sweeping a horizontal strip of sky. When she got back to where she started, she was to alter the elevation a little, and do the same again. And so on.

It is clear from Caroline's account that she was instructed rather than invited. "I found I was to be trained for an assistant Astronomer" (and indeed it was as an anonymous "assistant" that William would sometimes refer to her in correspondence). It is equally clear that at first she carried out William's bidding with little enthusiasm: "But it was not till the last two months of the same year before I felt the least encouragement for spending the starlight nights on a grass-plot covered with dew or hoar frost without a human being near enough to be within call." But she had always done as she was told; and this time the consequences would be momentous.

During her second night of sweeps, Caroline came across the nebula M27, that is, number 27 in the catalogue by the renowned French comet hunter Charles Messier that we still use today. As early as 1757 Messier had

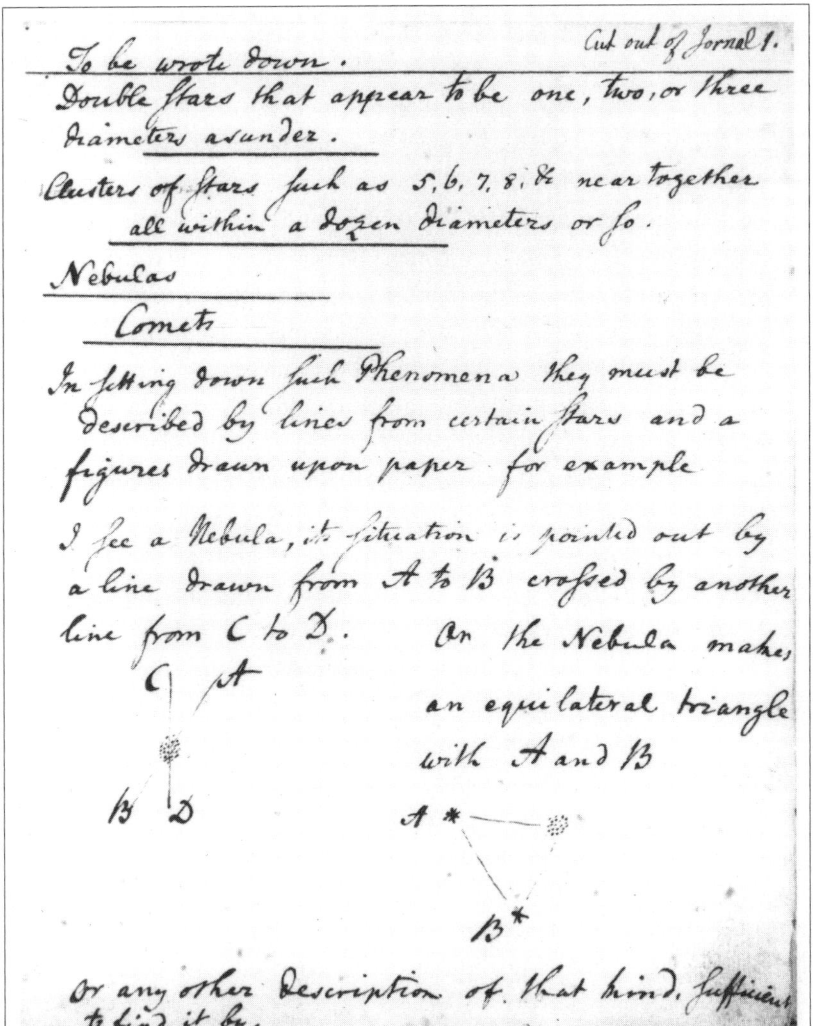

To be wrote down.

Cut out of Jornal 1.

Double stars that appear to be one, two, or three
diameters asunder.

Clusters of stars such as 5, 6, 7, 8, &c near together
all within a dozen diameters or so.

Nebulae

Comets

In setting down such Phenomena they must be
described by lines from certain stars and a
figures drawn upon paper. for example

I see a Nebula, its situation is pointed out by
a line drawn from A to B crossed by another
line from C to D.   Or the Nebula makes
                              an equilateral triangle
                              with A and B

Or any other Description of. that kind. Sufficient
to find it by.

Figure 8. Caroline's note of the instructions William gave her in the autumn of 1782 when she first began to search for objects of interest in the sky. RAS C.1/1.1, pasted in front, courtesy of the Royal Astronomical Society.

been employed to look out for the return of Halley's Comet, which was predicted for the winter of 1758–59, and the search for comets became his lifelong passion. It was while tracking another comet in the autumn of 1758 that he came across what we know as M1, the Crab Nebula. In September 1760 he came across a second nebula (our M2), and a third (M3) in May 1764. Less than a week later he found M4, which he could see was in fact a star cluster.

Nebulae and clusters looked like comets, and Messier would waste precious time, only to find that the "comet" he was supposedly tracking was in fact a permanent feature of the night sky. Accordingly, he decided to make a list of these objects, not because he was interested in them as such, but because they were proving a distraction in his searches for comets. His first list of 45 nebulae and clusters—some his own discoveries, some reported by other observers—was published in 1771. In 1780 Messier enlarged the list to 70 objects, and this appeared in the French reference annual *Connoissance des Temps* for 1783 (the *Connoissance* was published three years in advance to allow ample time for distribution). In the *Connoissance* for 1784 he enlarged the list again, to 103 objects, and this is the Messier catalogue that we use today.

Late in 1781 William Watson had learned that the *Connoissance* for 1783 listed "some nebulous stars." Not knowing that the list had already been superseded, he bought a copy and sent it to William as a present, and it was there that William and Caroline found M27. If the object was "known" to the astronomical community, it was new to William. In the weeks ahead only occasionally did Caroline find the opportunity to sweep, but as she did so she came across several more of Messier's nebulae, all but one new to William. Then, on February 26, 1783, she came across another nebula, and this time "My Brother examined it . . . Messier has it not" (figure 9).

Yes and no. The object was in fact the M93 of the final catalogue, but because William and Caroline were working with the 1780 version, it seemed to them that Caroline had made an entirely new discovery. Nor was this all, for a few minutes later she notes: "Following Gamma Canis Majoris a very faint nebula. . . . Messier has it not," and this time this was indeed the case. The vivid impression that William took away from the evening's viewing was that in a couple of hours his little sister, with a telescope that was little more than a toy, had added two completely new

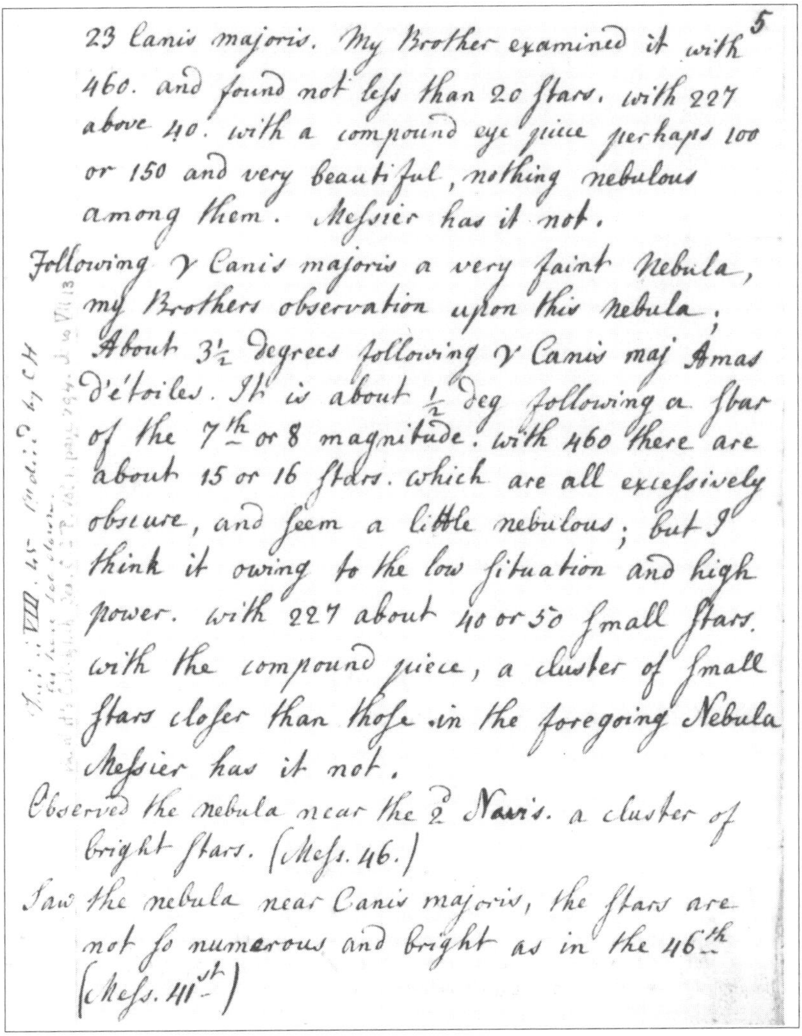

23 Canis majoris. My Brother examined it with 5
460. and found not less than 20 stars. with 227
above 40! with a compound eye piece perhaps 100
or 150 and very beautiful, nothing nebulous
among them. Messier has it not.
Following γ Canis majoris a very faint Nebula,
my Brothers observation upon this nebula.
About 3½ degrees following γ Canis maj. Amas
d'étoiles. It is about ½ deg following a star
of the 7th or 8 magnitude. with 460 there are
about 15 or 16 stars. which are all excessively
obscure, and seem a little nebulous; but I
think it owing to the low situation and high
power. with 227 about 40 or 50 small stars.
with the compound piece, a cluster of small
stars closer than those in the foregoing Nebula
Messier has it not.
Observed the nebula near the Q Navis. a cluster of
bright stars. (Mess. 46.)
Saw the nebula near Canis majoris, the stars are
not so numerous and bright as in the 46th
(Mess. 41st)

Figure 9. Part of Caroline's record of her "sweeps" on February 26, 1783. Of the
four nebulae she came across, the last two (M46 and M41) were in the 1780
catalogue of nebulae by Charles Messier that William Watson had sent William
as a present on December 7, 1781. But the first was not, and so Caroline wrote:
"Messier has it not." Unknown to William and Caroline, it was in fact M93 in
the enlarged catalogue Messier published in 1781, but it seemed to them that
Caroline had made a major discovery. When later the same night she came across
another nebula of which she could (rightly) say, "Messier has it not," William
became convinced that there were innumerable nebulae awaiting discovery. RAS
C.1/1.1, 5, courtesy of the Royal Astronomical Society.

nebulae to the sixty-eight so far known to science. In the realm of the nebulae there were indeed rich pickings to be had. Accordingly, on March 4 he broke off his search for double stars and "began to sweep the heaven for nebulas and clusters of stars."

A Revolution in the Making

These innocuous-sounding words were to have consequences for the history of astronomy whose importance it is impossible to exaggerate. Some (possibly all?) nebulae were distant clusters of stars, and the very existence of clusters proved that attractive forces were at work far beyond the solar system, pulling the stars toward each other. Now, some clusters were currently made up of widely scattered stars, while in others the stars were tightly packed. Over time, the stars in a scattered cluster were (it seemed) destined to move ever closer to each other, as a result of the continuing action of the forces that had formed the cluster in the first place. At the time of observation, therefore, such scattered clusters were young, while those whose stars were already tightly packed were at a later stage of their development: *clusters go through a life history*.

This was to prove an oversimplification. William later found that tightly packed clusters were globular in shape, and he understandably struggled to explain how this shape could come about. But this was a detail. His perception that clusters evolve tolled the death knell of the old astronomy according to which God the Clockmaker had made an unchanging creation whose components moved rather like the hands of a clock, and instead it ushered in our modern astronomy in which everything—individual stars, clusters, even the universe itself—has a life history. William and Caroline were about to lay the foundations for this fundamental transformation in astronomy, although its full realization would come about only when they were long gone.[1]

Nebulae had been discussed in antiquity by the Greek astronomer Ptolemy, but their modern history can be said to have begun in 1715 with a short paper in *Philosophical Transactions* by Edmond Halley. Halley had listed six nebulae, including that in Orion (M42), and had declared them to be

> luminous Spots or Patches, which discover themselves only by the Telescope, and appear to the naked Eye like small Fixt Stars; but in

reality are nothing else but the Light coming from an extraordinary great space in the Ether; through which a lucid *Medium* is diffused, that shines with its own proper Lustre.

William had first heard of these "lucid spots" in the sky from Smith's *Opticks*, and he learned more about them from Ferguson's *Astronomy*. When he opened his first observing book on March 1, 1774, the Orion Nebula was one of the two objects he examined (figure 10). That night the "seeing" was poor, too poor for William to make out the shape of the Orion Nebula, but on the fourth it was better. William knew what to expect, for Smith's *Opticks* provided the reader with a sketch of the nebula; although Smith did not say so, he had taken it from Huygens, and it gave a (misleading) impression of the appearance of the nebula in 1656. William was intrigued to find that the nebula in front of his eyes seemed to have altered shape since the time of the sketch. If so, it could hardly be a vast system of distant stars.[2]

He commented that its appearance was different from the drawing in Smith, sketched it himself, and remarked: "From this we may infer that there are undoubtedly changes among the fixt stars, and perhaps from a careful observation of this Spot something might be concluded concerning the Nature of it."

Coming at a time when professional astronomers had almost no interest in the universe that lay beyond the confines of our own little planetary system, William's insight is remarkable. If the Orion Nebula had changed shape in only a few decades—as William now had reason to think—then Halley must be right: this nebula was indeed a "lucid medium," a nearby cloud of glowing material (with embedded stars). True, a distant cluster of stars so far away that telescopes were not powerful enough to distinguish the component stars would also appear nebulous—and there were observers who believed that not only some but *all* nebulae were "unresolved" clusters of stars. But a vast star system could not radically change shape in only a matter of years. So was the Orion Nebula in fact changing shape, in which case it was without doubt a lucid medium, or was the apparent change an illusion?

William examined the Orion Nebula on no fewer than seventeen occasions during his Bath years, and he sketched its overall appearance three times. On most of these occasions he studied the well-defined layout of the embedded stars, and to help him detect any movements in these stars, he

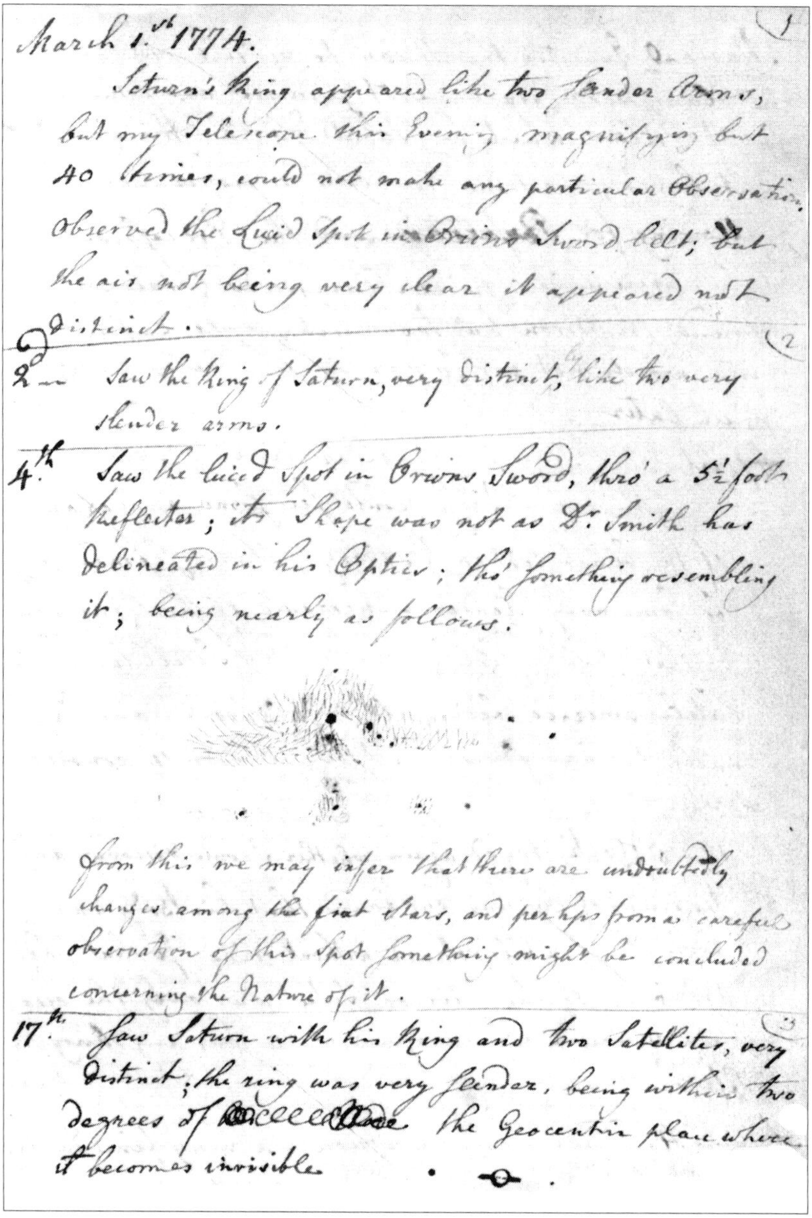

March 1st 1774.

Saturn's King appeared like two slender Arms, but my Telescope this Evening, magnifying but 40 times, could not make any particular Observation. Observed the Lucid Spot in Orions Sword belt; but the air not being very clear it appeared not distinct.

2. Saw the King of Saturn, very distinct, like two very slender arms.

4th Saw the lucid Spot in Orions Sword, thro' a 5½ foot Reflector; its Shape was not as Dr. Smith has delineated in his Opticks; tho' something resembling it; being nearly as follows.

From this we may infer that there are undoubtedly changes among the fixt Stars, and perhaps from a careful observation of this Spot something might be concluded concerning the Nature of it.

17th Saw Saturn with his Ring and two Satellites, very distinct; the ring was very slender, being within two degrees of ~~the~~ the Geocentric place where it becomes invisible.

Figure 10. The opening page of William's first observing book. On March 4, 1774, he observed and sketched the Orion Nebula and noted that its shape differed from that of the sketch (by Huygens) reproduced by Robert Smith in *Opticks*. If the nebula altered shape in only a matter of years, it could not be a vast and distant star system. RAS W.2/1.1, f. 1, courtesy of the Royal Astronomical Society.

made careful notes of which lay in straight lines, and so forth. His invariable conclusion was that the stars were motionless.

The "lucid part" was another matter. He was not yet experienced enough as an observer to appreciate that changes in the instrumentation, or in the level of moonlight, or in the "seeing" conditions, can generate spurious changes in an ill-defined object. And of course he had to rely on his own crude sketches made under difficult conditions. Nevertheless, on December 15, 1778, he decided "there is a visible alteration in the figure of the lucid part"; the following October, "The figure of the lucid part is very much alter'd."

In the late summer of 1779 he came across two more of Halley's nebulae. Twice he chanced on M13 in Hercules ("nebula with no stars in it"), and once on M11 in Scutum ("nebula seems to be a prodigious number of small Stars surrounded with luster and glare"). A year later, on August 6, 1780, he examined the Andromeda Nebula M31 and noted: "Has no star in it; but I believe I did not see the whole of it." Clearly, in the heavens there was a bewildering variety of nebulae. His curiosity piqued, with characteristic impetuosity he dedicated himself to their study: "To be observed. All the nebula's, their stars counted, and the form delineated."

The next night he looked again at M11 and M13. However, for the time being there was no obvious way in which he could fulfill his over-hasty commitment to study "all the nebula's" one by one; while by contrast his campaign to collect double stars was proving hugely productive. Occasionally he chanced upon a nebula not known to Smith or Ferguson. Then, in the summer of 1781, he looked again at M11 ("an amazing multitude of small stars," "great number of stars") and at M31 ("no star visible"). As before, he was baffled to know what to make of them.

More than a year was to pass before he saw his next nebula, and by then he would be a professional astronomer. This gap in time is surprising, for it was late in 1781 that Watson sent him Messier's 1780 catalogue.

In this catalogue there were riches indeed. In place of the tiny handful of nebulae that William had previously known about, he now had information on 70, with descriptions by the most experienced observer of the age. Some of the nebulae Messier declared to be clusters of stars, in particular M11, and William could confirm that this was so. Others were said to be apparently without stars, and this too matched William's own experience. But this might—or might not—be simply because the object was in fact a cluster of stars so distant that the small telescopes used by Messier in his hunt for comets had not been powerful enough to distinguish the individual stars. William however owned a 20-foot reflector with mirrors 12 inches in diameter; and this instrument might be able to "resolve" some of the Messier objects and reveal their true nature as star clusters—if indeed this was what they were.

William was not above telling a downright lie when it suited his purpose. By April 1784, when he sent to the Royal Society the first of his great papers on "the construction of the heavens," he had known of Messier's final list for nearly a year. He wrote:

> The excellent collection of nebulae and clusters of stars which has lately been given in the *Connoissance des Temps* for 1783 and 1784, leads me next to a subject which, indeed, must open a new view of the heavens. As soon as the first of these volumes came to my hands, I applied my former 20-feet reflector of 12 inches aperture to them; and saw, with the greatest pleasure, that most of the nebulae, which I had the opportunity of examining in proper situations, yielded to the force of my light and power, and were resolved into stars.

In fact it had been eight months after receiving the present from Watson that William had next looked at a nebula. On August 5, 1782, while on royal duty at Windsor Castle, he came across what he described in his observing book as "a Nebulous Star or Telescopic Comet"; for some time he was convinced he had found his first comet, but in the end the object proved to be M5. By August 20 the 20-foot was operational at Datchet, and it is only then that we find William seeking out Messier objects and examining them carefully. On the twenty-ninth he observed M52 and also M57, known today as the Ring Nebula in Lyra. The next night he was summoned to the castle, and he showed the king several nebulae, including M11 and M31: although William was still preoccupied with adding to his collection of double stars, his mind was beginning to turn to the problem of the nature of nebulae.

Early in September 1782 he examined M15 and M31, but he failed to find M26. Then, on September 7, he made his first major discovery among the nebulae, near the star Nu in Aquarius:

> A curious Nebula. or what else to call it I do not know. it is of a
> shape somewhat oval, nearly circular, and with this power [magnifi-
> cation of 460] appears to be about 10 or 15" diameter.

It had the near-circular outline of a planet, but the ghostly light of a nebula, and so he gave it the descriptive name of "planetary nebula." To check
whether it was a member of the solar system, he monitored its possible
movement with extreme care, but it proved to be as "fixed" as any of the
stars. He returned to the Saturn Nebula (as it is known today) time and
again, and visiting observers were regularly called upon to give their opinion of it; but it would continue to puzzle him throughout his career.

Two False Starts

Caroline meanwhile was devoting the limited time she could spare from
housework to her search for nebulae, and by the end of 1783 she reckoned
she had found fourteen. She even planned a catalogue.[3] It was a creditable
achievement, made possible by the ingenious Newtonian "sweeper" that
William designed and built for her that summer. Seated on a stool and
looking through the eyepiece at the upper end of the tube of the sweeper,
she would turn a handle and slowly rotate the tube from the horizontal
up to the zenith, examining a vertical strip of sky as she did so. She would
then move the telescope so that it faced in a slightly different direction
and rotate the tube again, this time from the zenith to the horizontal,
examining another strip of sky as she did. With practice she could sweep
the entire sky in only a few nights. But although William was proud to tell
his friends of what his sister was finding, individually her discoveries of
nebulae were of no consequence in the history of astronomy; some were
lost to sight until resurrected by the present writer, others were credited to
"C.H." in her brother's catalogues but would have been included in the
catalogues anyway.

But, as we have seen, overall they were of immense consequence because in March 1783 they inspired her brother to imitate her example.
There were to be two false starts before William's great search for nebulae
got properly under way. At first, in his impatience he overlooked the crucial difference for observers, between the nebulae (which are permanent

features of the night skies) and comets (which are temporary visitors to the solar system). The would-be discoverer of comets needed to make haste to search the heavens for a body that perhaps was not even there to be seen the previous night, for the mathematicians would be desperate to have as much information as possible about the orbit of any newly arriving comet. The student of nebulae, however, could take his time and use the biggest telescope at his disposal for a leisurely examination of these mysterious objects. William impulsively began to sweep for nebulae with a modest refractor before realizing his folly: the fine new 20-foot reflector he was building would be the ideal instrument for nebular research, and meanwhile he would be well advised, as a preliminary step, to familiarize himself with the prominent nebulae already listed by Messier. To study Messier objects William sometimes used his existing 20-foot; but observing from the top of a ladder in the dark was dangerous, and more often he used a 10-foot.

When the new 20-foot (the "large" 20-foot, because of its 18-inch mirrors; see plate 2) came into service in late October 1783, he made his second false start: he acted both as observer and as recorder of observations. Until now his only systematic observations had been of the brighter stars, which were easily visible in his telescope. And so when, for example, he had found that the Pole Star was double, he had used artificial light to make a written record of the fact, and then had been able to resume his viewing without delay. Because the next star to be examined would also be bright, he had not needed to wait (and lose valuable observing time) while his eyes became dark-adjusted once more. But the nebulae were among the faintest objects in the night sky, and if he used artificial light to write an account of some discovery, it would be minutes before his eyes were sufficiently dark-adjusted again for him to resume his search.

Not only that, but his method of maneuvering the reflector was ill thought-out. In its early months the ladder-type mounting that supported the great tube was fixed in a south-facing direction, and William would observe while standing on the nine-foot-wide platform and "swinging the suspended telescope backwards and forwards" by brute force, either side of south, through an arc of some twelve or fourteen degrees.[4] He would then alter the elevation a little and repeat the process; and again, and again. Physically it was exhausting work, because the tube was carried by central ropes and so had to be lifted as it was pulled to one side or the other. While all this was going on, the sky was rotating overhead, and so the region that had come under scrutiny and could be regarded as "swept" was ill defined.

The Natural History of the Nebulae

William was not slow to learn from his mistakes, and in mid-December he put in place a procedure that would underpin one of the greatest observing campaigns in the history of astronomy. He kept the tube of the reflector facing precisely south, in the manner of a transit instrument, and he relied on the rotation of the sky to bring new regions under his examination (plate 2). Because this rotation is slow, William found he had time to examine a horizontal strip of sky that was two degrees or more wide, several times the field of view of the instrument. He therefore recruited a workman whose job it was repeatedly to raise and lower the tube over the required arc, in a continual series of oscillations. To warn the workman when he had reached the limit in one direction and it was time to backtrack, Alexander devised a bell mechanism. After a few months William had succeeded in modifying the mounting so that the telescope could be rotated, and then, if an object of special interest came into view, sweeping would be interrupted while the object was followed for up to a quarter of an hour.

There remained the problem of the dark adjustment of his eyes. William simply could not afford to interrupt his examination of the sky and go into artificial light to make a written record of some new discovery. Caroline must do this for him. And so he installed Caroline at a desk at a nearby window, pen and paper in front of her, reference books to hand, and clock and dials alongside. The window would be closed to stop her freezing to death in the severe winters of the time; but when William pulled a cord as a signal, she would open the window, copy down his shouted account of the observation, give him any information he needed, and close the window again.

A French visitor who called in August 1784 describes her at work:

I arrived at Mr. Herschel's about ten o'clock. . . . Instead of the master of the house, I observed, in a window at the farther end of the room, a young lady seated at a table, which was surrounded with several lights; she had a large book open before her, a pen in her hand, and directed her attention alternately to the hands of a pendulum-clock, and the index of another instrument placed beside her, the use of which I did not know. She afterwards noted down her observations.

I approached softly on tiptoe, that I might not disturb a labour, which seemed to engage all the attention of her who was engaged in it; and, having got close behind her without being observed, I found that the book she consulted was the Astronomical Atlas of Flamsteed, and that, after looking at the indexes of both the instruments, she marked, upon a large manuscript chart, points which appeared to me to indicate stars.

This employment, the hour of the night, the youth of the fair student, and the profound silence which prevailed, interested me greatly. At last she turned round her head, accidentally, and discovered how much I was afraid to disturb her; she rose suddenly, and told me she was very sorry I had not informed her of my arrival, that she was engaged in following and recording the observations of her brother, who expected me, and who, in order that he might not lose the precious opportunity of so fine a night, was then busy in his observatory.

"My brother," said Miss Caroline Herschel, "has been studying these two hours; I do all I can to assist him here. That pendulum marks the time, and this instrument, the index of which communicates by strings with his telescopes, informs me, by signs which we have agreed upon, of whatever he observes. I mark upon that large chart the stars which he enumerates, or discovers in particular constellations, or even in the most distant parts of the sky." . . .

Placed at the upper end of his telescope, when the indefatigable astronomer discovers in the most deserted parts of the sky a nebula, or a star of the least magnitude, invisible to the naked eye, he informs his sister of it, by means of a string which communicates with the room where she sits; upon the signal being given, the sister opens the window, and the brother asks her whatever information he wants. Miss Caroline Herschel, after consulting the manuscript tables before her, replies, brother, search near the star *Gamma, Orion*, or any other constellation which she has occasion to name. She then shuts the window, and returns to her employment.[5]

The next day she would write up a fair copy of the night's work. And when they had accumulated one thousand new nebulae and clusters of stars, she would prepare a catalogue to be sent to the Royal Society for publication. Caroline was William's photocopier, word processor, and calculator rolled into one.

Of course it was Caroline who kept a tally of which areas of sky had been swept and which not. She took large sheets of paper and drew on them horizontal and vertical lines, each little square representing an area of sky measuring fifteen minutes in each coordinate. A cross within a square indicated that it had been carefully swept, a diagonal that it had been seen but needed to be reexamined.[6]

The campaign was fully under way when the Portuguese-born astronomer Jean-Hyacinthe Magellan called early in 1785.

I spent the night of the 6th January at Herschel's, in Datchet, near Windsor, and had the good luck to hit on a fine evening. He has his twenty foot Newtonian telescope in the open air and mounted in his garden very simply and conveniently. It is moved by an assistant who stands below it. . . . Near the instrument is a clock regulated to sidereal time. . . . In the room near it sits Herschel's sister and she has Flamsteed's Atlas open before her. As he gives her the word, she writes down the declination and right ascension and the other circumstances of the observation. . . . I went to bed about one o'clock, and up to that time he had found that night four or five new nebulae. The thermometer in the garden stood at 13° Fahrenheit; but in spite of this, Herschel observes the whole night through, except that he stops every three or four hours and goes into the room for a few moments. For some years Herschel has observed the heavens every hour when the weather is clear, and this always in the open air, because he says that the telescope only performs well when it is at the same temperature as the air. He protects himself against the weather by putting on more clothing. He has an excellent constitution and thinks about nothing else in the world but the celestial bodies.[7]

Caroline's recruitment as William's amanuensis put a stop for the time being to her own work as an observer. This was a huge relief to her. For

William it went against the grain to have a major telescope standing idle, and so once he had commissioned the new 20-foot, back in October, he had assigned its predecessor to the unfortunate Caroline. In the most bizarre episode of their entire half-century partnership, she was told to climb the ladder and reexamine William's double stars to see if any members of the pairs had altered position with respect to one another in the interval since their discovery. Whereas the observer at the new reflector stood securely on a railed platform or, if he preferred, could sit comfortably in an adjustable chair attached to the structure supporting the great tube, at its predecessor the observer was precariously perched on a ladder as much as twenty feet above the ground. For a lady in a dress, to climb this ladder in the dark and then, while clinging to the upper rungs, to try to locate and measure a double star was a frightening experience with zero chance of success. The whole episode is so preposterous that if we did not have an account of it in Caroline's own hand, we might dismiss it as fantasy.

In their new (and eminently sensible) collaboration, whenever he came across a nebula, William would report its position to Caroline by reference to a nearby star, on the lines of "up by so much, left by so much," and he relied on Caroline to identify the star in question. The clock gave her the position of the star in one coordinate, while the "other instrument" mentioned by their French visitor told her the altitude of the telescope, to which it was connected by a cord, and so gave the other coordinate.

There was however a minor problem. John Flamsteed's British Catalogue of stars was the bible of observers; but it was organized by constellation, and as William was observing at the telescope, the rotation of the sky might well take him from one constellation to another. This required Caroline suddenly to turn the pages of the British Catalogue from the former constellation to the latter, with all the attendant disruption. The resourceful scribe therefore compiled from Flamsteed a catalogue that exactly met her requirements (figure 11). On a given night, all the objects that William would examine would be at much the same angular distance from the celestial North Pole. Caroline therefore organized a catalogue of stars arranged in five-degree "zones" of North Polar Distance, and then in the order in which the rotation of the sky would bring them to William's inspection. She could now forewarn her brother of the stars that would next come into view, and when he defined the position of a nebula by reference to a nearby star, she would know which star it was. The first

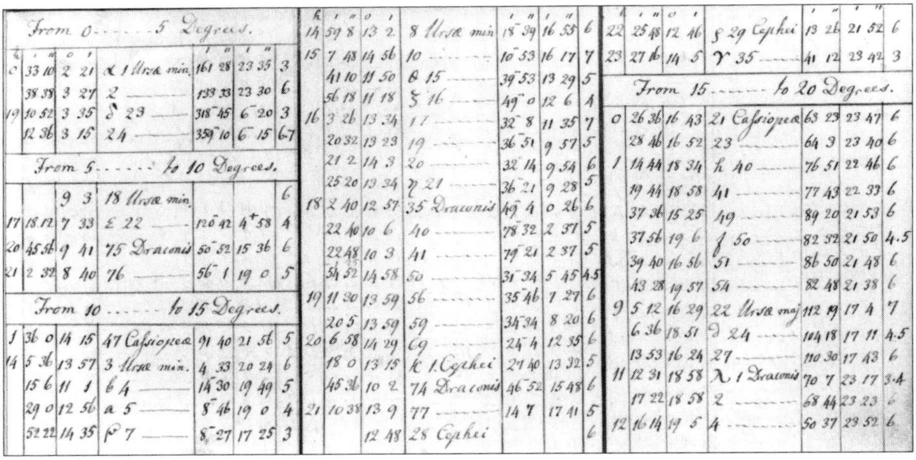

Figure 11. The first page of Caroline's 1786 list of stars, which she had in front of her while William swept for nebulae and clusters. In Flamsteed's British Catalogue the stars were arranged by constellation, and this was most inconvenient because William's gaze would cross the constellation borders as he swept. Caroline knew that all the stars in a given sweep would be at roughly the same angular distance from the pole. She therefore prepared her own list of the brighter stars in the British Catalogue, arranged in "zones" according to their angular distance from the pole and then in the order (expressed in time) in which they would present themselves to William. The first two columns give the order (adapted from Flamsteed), the third the distance from the pole according to Flamsteed, the fourth the name of the star, the fifth and sixth the changes in the coordinates since Flamsteed's day, while the final column gives the magnitude of the star (with magnitude 6 the faintest visible to the naked eye). It is impossible to exaggerate the value to William of Caroline's meticulous work as his amanuensis, of which this is but one modest example. RAS C.2/1.2, courtesy of the Royal Astronomical Society.

version of this catalogue did not extend to stars high in the sky, "The apparatus not being then ready for sweeping in the zenith," but by 1786 the catalogue was complete. The collaboration between William and Caroline, with gadgetry devised by Alexander as required, would prove to be astonishingly productive, and by the time they called a halt in 1802, they had added no fewer than 2,510 nebulae and clusters to the hundred or so listed by Messier.

Caroline now had a purpose in life, indeed several. She managed the

Herschel household, she was William's indispensable assistant in a great scientific enterprise, she had her own research program to pursue with her ingenious sweeper when her brother did not need her, and she was endlessly useful to him in all manner of ways. Sometimes she paid a price for her devotion. So, for example, only a few days after the collaborative search for nebulae had got under way, the skies cleared unexpectedly about ten o'clock, and there was a rush to ready the 20-foot and begin observations. William shouted down to tell Caroline to make an adjustment in the machinery, and as she ran in the dark in melting snow, she impaled her thigh on an iron hook.

> [M]y brothers call[,] make haste[!] I could only answer by a pittiful crey[,] I am hooked. He and the workman were instantly with me, but they could not lift me without leaving nearly 2 oz. of my flesh behind.

She was out of action for a while—her doctor said that a soldier with such an injury would have been entitled to six weeks' nursing—and it is revealing that she comments without a trace of irony:

> I had, however, the comfort to know that my Brother was no loser through this accident for the remainder of the night was cloudy and several nights afterwards afforded only a few short intervals favourable for sweeping.

The Riddle of the Milky Way

Only a matter of weeks after beginning his sweeps for nebulae, William sent to the Royal Society the first of his great series of papers on the construction of the heavens, papers through which cosmology lost its midcentury status as a mere playground for amateur speculators, and became an authentic if still-embryonic branch of astronomy. His new 20-foot reflector was a masterpiece; his recently inaugurated sweeps for nebulae and clusters would prove to be one of the most remarkable observational campaigns in history; now it was time to begin to theorize about the insights he had gained.

The paper is dated April 1784, and by the time it was dispatched,

William had already discovered 388 nebulae unknown to Messier. By June 17, when the paper was formally "read" to the Royal Society, the number had risen to 440, and by the time he corrected proofs, he had 466 in the bag.

William was understandably keen to win credit for his discoveries. Their sheer quantity was impressive, but to demonstrate the quality of his observations, he proudly lists Messier's unresolved "nebulae" that his powerful new reflector has shown in fact to be star clusters. As a result it is easy to read him as saying that *all* nebulae are star clusters, but this is not his meaning. In Bath he had, he believed, detected changes in the shape of the Orion Nebula; and this had since been (supposedly) confirmed by his observations in the summer of 1783, before the completion of the new 20-foot. On September 20 of that year the nebula was again "changed," and on the twenty-eighth "surprisingly changed." It must therefore be a true nebula, small and close to Earth, and not a vast star cluster at a distance so great that it appeared nebulous; and so, when he examined another of Messier's nebulae that August, he noted that he could not decide whether this object was a cluster or a true nebula "such as in Orion's Sword handle."

But these thoughts about the nature of the nebulae were only the start of what his paper had to offer readers of *Philosophical Transactions*. William had turned his reflector to the Milky Way, and he confirmed that it was indeed formed of myriad stars. He counted those visible in his telescope in sample directions and concluded that in an hour's observing he might see fifty thousand that were individually distinct, and glimpse many more. Furthermore, in his sweeps for nebulae and clusters he had been struck by how these were to be found in concentrations in certain parts of the sky: "they are arranged into strata, which seem to run on to a great length." Combining these two insights in a brilliant leap of imagination, he declared that the Milky Way itself "undoubtedly is nothing but a stratum of fixed stars" not yet fragmented into nebulae and clusters.

This would explain the optical effect that we see in the night sky. For if the Earth is indeed immersed in a stratum of stars, then as we look out in directions away from the stratum, we shall see only a few near (and therefore bright) stars before our gaze reaches out into empty space; whereas if we look around us in directions within the stratum we shall see great numbers of stars near and far, bright and faint, whose light will merge to give the appearance of milkiness. In other words, the stratum model explains the appearance of the Milky Way.

William was not the first to think along these lines. In 1750 the religious speculator Thomas Wright of Durham (to distinguish him from other writers of the same name) published a book in which he argued that the stars of the system to which the Sun belongs occupy a spherical shell of space, in the midst of which is a Divine Center. The shell, he says, has a vast diameter, so vast that the curvature of the shell is barely perceptible in the stars that are actually visible to us. The stars we see, therefore, can for all practical purposes be thought of as sandwiched between two planes, one "plane" being in fact part of the inner surface of the enormous shell and the other, the outer.

William at some time acquired a copy of Wright's very confused book, although there is little sign that he actually read it. But it is not impossible that, during his time in the north of England, William heard mention of Wright's concept and then forgot about it until it surfaced when he reflected on the problem of the Milky Way. A brief account of Wright's book had earlier come to the eyes of the German philosopher Immanuel Kant, who misunderstood what Wright was trying to say and thought he was arguing that the stars of our system do indeed form a flat stratum. But Kant's book on the subject had limited circulation because of the bankruptcy of the bookseller, and it is very unlikely that it ever came to William's notice. A third speculator along these lines was the Alsatian polymath Johann Heinrich Lambert, but Lambert's book did not come into William's possession until 1799. His reaction then was to pen a scathingly hostile critique that ran to ten foolscap pages. It seems most likely, therefore, that William arrived at the concept independently, as a result of his reflections on the already-fragmented "strata" of nebulae and clusters that he believed he was encountering.

The Abandonment of "True Nebulosity"

Since William currently believed that some nebulae, such as that in Orion's Sword, were formed of true nebulosity ("gaseous," as we would say), while others were star clusters at great distances, he was faced with the problem of how to distinguish between the two. His solution was plausible: some nebulae "contain a nebulosity of the milky kind, like that wonderful, inexplicable phaenomenon about Theta Orionis [the Orion Nebula]; while others shine with a fainter, mottled kind of light, which denotes their being resolvable into stars." That is, there were two kinds of nebulous appearance: the smooth, milky nebulosity of a true nebula and the irregular, mottled appearance of a distant star cluster.

As luck would have it, just five days after the paper was read to the Royal Society, William came across M17, today known as the Omega Nebula. In it, both types of nebulosity seemed to coexist:

> A wonderful Nebula. . . . It is not of equal brightness throughout, and has one or more places, where the milky nebulosity seems to degenerate into the resolvable kind. . . . Should this be confirmed on a very fine night, it would bring on the step between these two nebulosities which is at present wanting, and would lead us to surmise that this nebula is a stupendous Stratum of immensely distant fixed stars some of whose branches are near enough to us to be visible as resolvable nebulosity, while the rest runs on to so great a distance as only to appear under the milky form.

In other words, he had drastically revised his interpretation of the two nebulosities: mottled nebulosity now indicated the presence of stars in the middle distance, while milky nebulosity was the collected light of stars at very great distances. In consequence, "true nebulosity" was no more than a figment of the imagination; and so the changes he believed he had detected in the Orion Nebula must be illusory.

William deserves credit for having the flexibility of mind to abandon his earlier theory in the face of new evidence. And he found confirmation (or so it seemed) of his new opinion four weeks later, when his sweeps brought him once more to M27, the Dumbbell Nebula. Caroline had seen it on her second night of sweeping with her little refractor, but now it came under examination with a major telescope. William thought he detected not only the milky nebulosity of the stars in its most distant regions, together with the mottled nebulosity of stars in the middle distance, but even the individual stars of the region of the nebula closest to us. It was, he decided, a stratum of stars—more exactly, a double stratum—seen by us edge-on.

William's next paper on the construction of the heavens was read to the Royal Society in February 1785. A universe of stars without nebulosity was a much simpler one for his reader to grasp, and this enabled him to argue clearly. He opens the paper with a remarkable declaration, that in his opinion it is better to speculate too much than too little. Not every author who publishes in a scientific journal is so frank. Of course, he says, one must avoid indulging "a fanciful imagination." But

> if we add observation to observation, without attempting to draw not only certain conclusions, but also conjectural views from them, we offend against the very end for which only observations ought to be made. I will endeavour to keep a proper medium; but if I should deviate from that, I could wish not to fall into the latter error.

The very existence of clusters (he argues) demonstrates that attraction is at work among the stars. Imagine, then, an early stage in the history of the universe when there are "numberless stars of various sizes, scattered over an indefinite portion of space in such a manner as to be almost equally distributed throughout the whole." Although William did not know it, this was the model of the universe—but with "infinite" rather than the vaguer "indefinite"—that Isaac Newton had discussed a century earlier with his most intimate friends. Both men asked themselves what would happen in such a universe of stars when gravity got to work.

When Newton was in his prime, the first proper motions of stars had yet to be identified. The Greeks had believed the stars to be "fixed," and after two millennia they were seemingly as fixed as ever. This was a problem for Newton, for he believed both that the stars were isolated bodies free to move in any direction and that the law of attraction—gravity—operated throughout the universe. But if every star pulled every other star, how was it that they were all seemingly at rest? The short-term answer, he decided, was that God's Providence had created the universe of stars with near-perfect symmetry, so that each star was pulled almost equally in opposite directions by the combined attractions of all the other stars. And he analyzed the numbers of stars listed in the catalogues to offer some justification for this claim of symmetry, at least for our region of the universe.

Curiously, Newton did not see the Milky Way for what it is, evident proof that the system of the stars is far from symmetric. But even he had to admit that the symmetry he claimed was not perfect, and therefore it was only a matter of time before the stars would begin to move from their appointed places. But God was the great Clockmaker, and He had what we might think of as a regular servicing contract with His universe: He would intervene when problems threatened and move the stars back to where they should be. This was no panic-miracle on God's part, not the response to an emergency call, but rather the way in which He had always intended to demonstrate His continuing loving concern for the men and women He had created.

William, by contrast, saw himself as living in a cosmos that changes, evolves. He too begins with an imagined near-perfect symmetry in the universe of stars. And like Newton, he accepts that "the indefinite extent of the sidereal heavens . . . must produce a balance that will effectually secure all the great parts of the whole from approaching to each other." But, unlike Newton, he accepts that because the symmetry is not quite perfect, and therefore the gravitational pulls will be greater in some places than in others, the system of stars will eventually fragment. The greater attractive force that results from the presence of an unusually large star, or of an exceptional concentration of normal stars, will pull in the surrounding ones, leaving empty spaces behind (figure 12). Eventually clusters of stars will form; and within each cluster attraction will continue its remorseless operation, and perhaps in the long run the cluster will experience what we would call gravitational collapse. But unlike Newton, William sees "the destruction now and then of a star, in some thousands of ages, as perhaps the very means by which the whole is preserved and renewed."

> These clusters may be the *Laboratories* of the universe, if I may so express myself, wherein the most salutary remedies for the decay of the whole are prepared.

Fathoming the Milky Way

William speculates about the types of star systems that might result from the various possible irregularities in the initial near-symmetric distribution of stars, and he believes that our own Milky Way system is an example of his third type; it is "A very extensive, branching, compound Congeries of many millions of stars." We are inside this "Congeries," but in spite of this handicap he believes that we can plot its outline. To do this he makes two assumptions. The first is that his 20-foot reflector is able reach the border of the Milky Way in all directions. He cannot prove that this is so, but

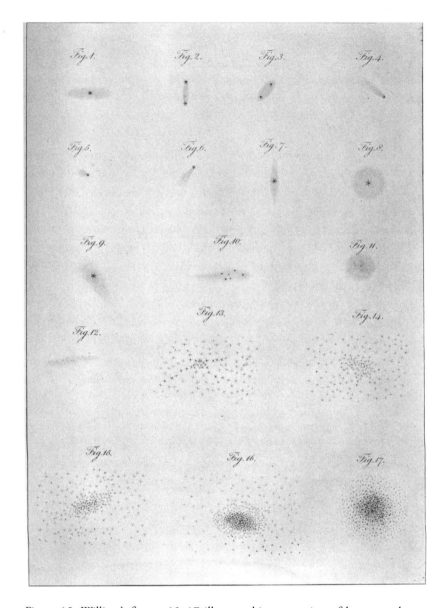

Figure 12. William's figures 13–17 illustrate his conception of how star clusters increasingly condense over time under the action of gravity. This sketch comes from his paper in *Philosophical Transactions* in 1814 on "the sidereal part of the heavens" (when he once again believed in "true nebulosity") but he arrived at the concept in the 1780s.

unless he is allowed this assumption, his enterprise is obviously doomed from the start.

The second assumption is more controversial. He had supposed in his theoretical discussion that the universe of stars began with a near-uniform distribution, and he now assumes that, despite the changes wrought by gravity since then, there is still a reasonably uniform distribution within the borders of the Milky Way. This being so, the number of stars he could see in a given field of view of his reflector was directly related to the distance to the border in that direction: the more stars, the further the Milky Way extended in that direction.

William could not spare the time to count stars over the whole sky, but he showed how things were to be done by going around a "great circle" that would represent a cross section of the Milky Way (figure 13). His paper contains a table of nearly 700 "star-gages," and most of these are the average of ten counts of neighboring fields. It was the first great example in history of stellar statistics, a tool now in everyday use among astronomers. As expected, where his great circle crossed the plane of the Milky Way, the counts were exceptionally high, which confirmed his opinion that the Milky Way system is a broad "stratum" of stars.

That it was still a stratum and had not yet fragmented encouraged William to see it as at an early stage of development, with "a certain air of youth and vigour"—although, he says, there are many places in the Milky Way where the stars "are now drawing towards various secondary centers, and will in time separate into different clusters." In his sweeps he had several times encountered regions outside the Milky Way where there was a collection of such clusters, and each of these collections he now saw as the separated fragments of a stratum that in the past had been similar to the Milky Way.

The Orion Nebula he now believed to be, like all nebulae, a system of stars, and it was so distant that he had not yet succeeded in resolving it into its component stars. Since despite its distance it appeared so large, its diameter measured in miles must be enormous, so much so that it "may well outvie our milky-way in grandeur." The same was true of other prominent nebulae, notably the one in Andromeda. All these he believed to be what we would term "galaxies."

William ends this astonishing paper, among the greatest in the whole

Fig. 4.

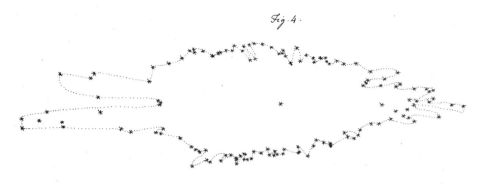

Figure 13. A preliminary sketch for the cross section of Milky Way published by William in 1785 in his paper in *Philosophical Transactions*, "On the Construction of the Heavens." The cross section was based on the assumptions that his 20-foot reflector could reach the borders of the Milky Way in all directions and that within the Milky Way the distribution of the stars was roughly uniform. He later abandoned both assumptions, but astronomers abhor a vacuum, and the diagram was being reprinted in textbooks long after William was dead. RAS W.4/24.1, courtesy of the Royal Astronomical Society.

history of astronomy, with a discussion of his planetary nebulae, beginning with the one near the star Nu Aquarii, which had puzzled him so much. He considers a variety of explanations, none of which he finds convincing, and finally concludes that the planetaries are most likely to be tightly packed clusters of stars:

> If it were not perhaps too hazardous to pursue a former surmise of a renewal in what I figuratively called the Laboratories of the universe, the stars forming these extraordinary nebulae, by some decay or waste of nature, . . . may rush at last together, and either in succession, or by one general tremendous shock, unite in a new body.

Perhaps the new star that had appeared in Tycho Brahe's time was such an implosion.

William would one day abandon several of the hypotheses that underlay this paper. He came to recognize that not all nebulae are star clusters, that his 20-foot was not able to penetrate to the borders of the Milky Way, and that high counts of stars in the Milky Way reflected clustering rather than distance to the border.[8] And yet if any single paper marks the opening of the modern era in astronomy, this is it. In place of Newton's clockwork universe in which God intervenes when attraction threatens its stability, we have a cosmos in which attraction is the all-embracing agent of change. Attraction exploits irregularities in the initial distribution of stars to bring about changes that lead to strata like our Milky Way, and then to fragmentation of these strata into separate clusters, with eventual gravitational collapse leading to a renewal that William does not claim as yet to understand. As to the Milky Way itself, what we see in the sky is the optical effect of our immersion in a stratum, and its outlines we can determine by the simple expedient of counting stars. And this stratum we inhabit is only one among many; it is big, but others are even bigger.

When William's paper was read to the Royal Society, only a little over two years had elapsed since King George invited William to become astronomer to the Court at Windsor. In this time William the instrument maker had built the "large" 20-foot reflector, one of the great telescopes of all time. William the observer had published a second catalogue of 434 double stars and had already increased the number of nebulae and clusters known to science by a factor of nine. William the theoretician had determined the solar apex, and he had opened up the cosmos to scientific investigation with his discussion of the evolution of star clusters, his measurement of the shape of our Milky Way galaxy, and his identification of comparable galaxies elsewhere in the universe. All this and much more for £200 a year. But between William and King George there was trouble in store.

7

1782–1790
"One of the Greatest Mechanics of his Day"

Moonlighting by Daytime: Telescopes for Sale

Caroline's anxiety as to how they would manage in Datchet on £200 a year would have grown to little short of panic if she had realized how much the wide-aperture telescopes that William was planning would cost, for only reflectors with great "light gathering power" could bring into view the distant and faint components of the construction of the heavens. The large 20-foot was commissioned a year after their arrival at their new home, and William had paid every penny out of the savings he had accumulated in Bath. But the Herschel household could not sustain a negative cash flow indefinitely.

From the start of his own career as an observer, William had tried to persuade his siblings to share his enthusiasm, and at Bath he had made small reflectors of the user-friendly "Gregorian" type for Caroline, Alexander, and Dietrich—although whether any of them ever observed while in Bath is open to doubt. And he had occasionally manufactured instruments to satisfy local friends, William Watson in particular. Now European astronomers were waking up to the fact that William's homemade reflectors were without equal, and as early as August 1782 Christian Mayer of Mannheim had written to ask for a Herschel telescope, money no object. William's declined, but the wording of his reply suggests that Mayer's approach had opened his mind to the thought of commercial profit:

> I made telescopes only to indulge the very great love I had for astronomical observations, therefore it will at present not be possible to furnish you with such a one; but I believe in a short time I shall

give the model of my Instruments and the method of making the Specula [mirrors] to some Optician, and if that is found practicable, I suppose one of my 7 feet Telescopes might be made for about 50 guineas; but I fear opticians will hardly bestow that excessive care and attention which are necessary to form a truly parabolic Speculum of a considerable diameter.[1]

A year later, soon after the commissioning of the large 20-foot, King George found himself wondering whether he had driven too hard a bargain with his astronomer. William had ambitions that should be encouraged, and the king had no wish to see him reduced to penury. He therefore suggested that William use his spare time to develop a commercial business manufacturing telescopes, and George himself launched the new enterprise with an order for five 10-foot reflectors. There could be no question of William's entrusting the crucial optical components of these royal reflectors to any tradesman: the mirrors and eyepieces he must make with his own hands. But the wooden mountings and the brass work could be contracted out.

It was an act of generosity on the part of the king, or perhaps the promptings of a guilty conscience, for he had not even thought of what he would do with the fourth and fifth instruments. Windsor Castle and Kew Observatory were obvious destinations, as was Göttingen University, a favorite good cause of George's in his capacity as Elector of Hanover. The fourth instrument (plate 3) the king eventually presented in 1786 to the Duke of Marlborough to thank him for his hospitality during a royal visit to Blenheim Palace. But as late as 1791 we find William writing to George to point out that the fifth 10-foot was still without a home (and to remind the king that his astronomer was awaiting payment for all five).

By the spring of 1785 the production of Herschel telescopes was in full swing. William tells us that he "soon found a great demand for 7 feet reflectors. This business, in the end, not only proved very lucrative but also enabled me to make expensive experiments for polishing mirrors by machinery." He wrote to Alexander:

I have bought a compleat set of tools for working in Brass; erected a small forge; have a Brass workman; the Cabinet maker is imployed; the Joiner at work; the Smith forging away; so that I hope to get some instruments finished pretty fast.

Alexander himself would come and help with the brass work in the summer, when Bath was out of season, and he proved so adept that William
tried to set him up as a telescope maker on his own account. But Alexander
preferred to work alongside William, and after he became a widower in
1788, he would regularly spend time with his brother when his musical
duties allowed. The journey was simple, for the London coach stopped
almost outside the home in Slough where William lived from 1786, and
this allowed Alexander to bring his tools with him:

> [T]his eccentric man . . . was a musician by profession, but all his
> thoughts sleeping or waking were directed to the subject of mechan-
> ics. He never moved away from his own home, except to pay a yearly
> visit to his brother's family, and then invariably came accompanied
> by his turning-lathe and other implements, and getting himself &
> them established the moment of his arrival, in the workshop (now
> H's observatory) scarcely left that apartment during the whole pe-
> riod of his stay. His appearance at meals with the family was never
> thought of—and he was indulged by them in his humour to his
> heart's content—allowed to pass his time in his own way & never
> asked one question or interrupted in his pursuits by any thing. He
> used to go away after his week of visitation had expired, having
> scarcely seen his friends all the time, but declaring himself quite
> delighted with their society.[2]

The Greatest Reflector Ever Built

William (plate 4) hoped to contrive a method of polishing mirrors by
machinery because he had revived his Bath ambition to build a reflector
with 3- or even 4-foot mirrors: a telescope of this size might settle, once
and for all, the great question of whether all nebulae were nothing more
than distant star clusters.[3] Such mirrors would require a huge quantity of
metal, not least because they would need to be thick enough to keep their
shape when tilted in the tube. The disks would have to be cast in London
and shipped upriver to Windsor, transported overland to William's home,
and then ground into parabolic shape and polished.

The expense would be horrendous. And the mirror was only the start.
The wooden mounting and observing platform of his new 20-foot had

proved outstandingly successful, but to scale it up to monster dimensions would demand a vast input of labor and materials. And if all went well there would then be the constant expense of keeping the instrument in commission. To build and to maintain, the great reflector would cost a king's ransom—and the only way to fund it was to hold King George to ransom.

One of the visitors to Datchet had been William's old ally, William Watson. Watson had been outraged when William told him in confidence the amount of his pension, and he was now further outraged to find it was not as obvious to everyone else as it was to him, that William must be given whatever money he needed for the magnum opus he now planned. Watson

> saw my Brothers difficulties and expressed great dissatisfaction. And on his return to Bath he met among the Visitors there several belonging to the Court . . . to whom he gave his opinion concerning his Friend and his situation very freely.

The Datchet house was uncomfortably close to the river Thames. William, "when the waters were out round his garden, used to rub himself all over, face and hands &c., with a raw onion, to keep off the infection of the ague" as he observed in the bitter cold. And still rain came through the ceilings. In June 1785, therefore, they moved across the river to Clay Hall in the village of Old Windsor (figure 14), within sight of the castle. They swept at Datchet for the last time on May 30, and for the first time at Clay Hall on June 10. A couple of months later William successfully petitioned the king for a grant of £2,000, representing the estimated capital cost of the new telescope plus the running costs for four years.[4]

The Move to Slough

By September 1785 the tools for grinding and polishing had been prepared, and the mirror was cast in London at the end of October. A problem then arose. Clay Hall had been inherited the previous year by Mrs. Laura Keppel, the illegitimate daughter of Sir Edward Walpole and the widow of the Bishop of Exeter. Laura was described by Horace Walpole as a "fiery furnace," and William found her "a litigious woman" who raised the rent every time he made an improvement to the property. The large

Figure 14. Clay Hall in Old Windsor, where the Herschels lived from June 1785 to March 1786. Herschel Family Archives.

20-foot had been moved from Datchet and could be moved again, but the 40-foot reflector would need a permanent home. The Herschels were friends of Frederic Albert, one of the king's pages, who rented a small ivy-covered house in Windsor Road, Slough; it was said to have once been a tavern—certainly it had ample beer cellars, an asset at a time when water was often unfit to drink. Its white-painted porch led straight from the road into the hall, and upstairs there were four bedrooms. The house had stables and outhouses and a pretty walled garden of one acre, at the end of which was a gravel walk, with a row on high elms on either side.[5] It was surrounded by cornfields and pastures that sloped southward, down to the Thames flood meadows at Eton, beyond which rose the majestic outline of Windsor Castle.[6] It would be ideal for William's purposes, except for the trees that obscured the horizon; but these could be felled.

Albert's three-year lease was to end the following spring on March 25[7] (Lady Day, the customary date for the transfer of a property), and William and Caroline arranged to rent the house (figure 15). They swept at Clay Hall for the last time on March 28, and the following day we find William writing letters from Slough; on April 4 Watson arrived there in the hope that the move was already complete enough for him to have a bed for the

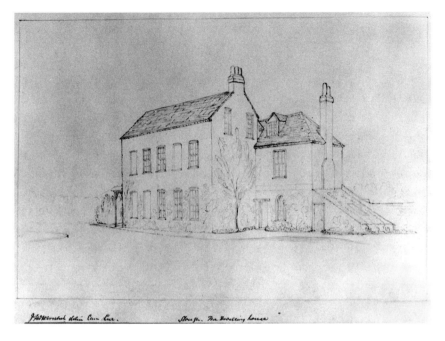

Figure 15. The Herschel home in Slough, "the place in the world where the most discoveries have been made," according to the French astronomer François Arago. Drawing by John Herschel with a camera lucida, Herschel Family Archives.

night. The 20-foot had been erected at Slough on the third, and that evening William and Caroline had made a first, token sweep (figure 16); but William still had to determine the direction of due south, and the clocks had been disturbed by the move, and so it would be a couple of weeks before serious work could resume.

The Grove, which survived as Observatory House until August 1960, was to be William's home for the rest of his life. The French astronomer François Arago would one day describe it as "the place in the world where the most discoveries have been made";[8] because of what William had achieved there, Arago added, the very name of Slough would be handed down "religieusement" by astronomers from generation to generation. In England, the words "Slough" and "astronomy" would become linked in the public mind (figure 17). But Caroline took a less romantic view. She was to remember the "swarm of pilfering work-people, with which Slough, I believe, was particularly infested."

Caroline's reservations notwithstanding, the move to Slough was nothing short of providential. Datchet and Old Windsor had been isolated villages, whereas Slough lay astride the Great West Road from London to Bath, at the point where it was crossed by the road from Windsor heading north. Coaching for the wealthy few had developed in England in the early seventeenth century, and as journeys could take several days, there was need for inns where the travelers could stay overnight. In 1618 a local squire had built two coaching inns there, the Crown and the Reindeer, and by midcentury, when the first scheduled stagecoaches passed through Slough on the three-day journey from London to Bath, these had been joined by two more.[9] In 1701 William Baldwin was granted a long lease on the Crown.[10] A decade or so later, when the journey time from London to Bath had been reduced to two days, his enterprising son Thomas, "a citizen and cooper of London,"[11] instituted the first daily coach service between London and Bath, and "held the road for many years against all rivals."[12] Fortunately for the Baldwins, the period that followed saw the transformation of Bath from a quiet market town with medicinal springs into the most fashionable center outside London itself. Royalty and aristocracy flocked there to take the waters and enjoy the many entertainments, and those who provided transport from London, and the coaching inns along the route, prospered.

In 1743 the upgrading of the London-to-Bath road to a "turnpike" was completed (the term referred to the barriers that ensured a coach would pay the appropriate contribution supposedly for the upkeep of the highway), and at the overnight stop at Newbury, there were nine inns, including the George & Pelican, which could accommodate no fewer than three hundred horses.[13] Coaches by the dozen passed each day along the road, and it was a simple matter for Alexander to travel from Bath to Slough to help William with his brass work, or for William to go up to town for a meeting of the Royal Society, or for the society's president to visit William to view the latest comet.

The development of the railway would one day bring coach travel to an end, but in the mid-eighteenth century there were fortunes to be made from the Great West Road, and the Baldwins made them. And although he could not know it, William's future mother-in-law was his current landlady, Mrs. Elizabeth Baldwin.

Work began on the stand of the 40-foot soon after the move to Slough. Unfortunately the king proved to have no conception of what was

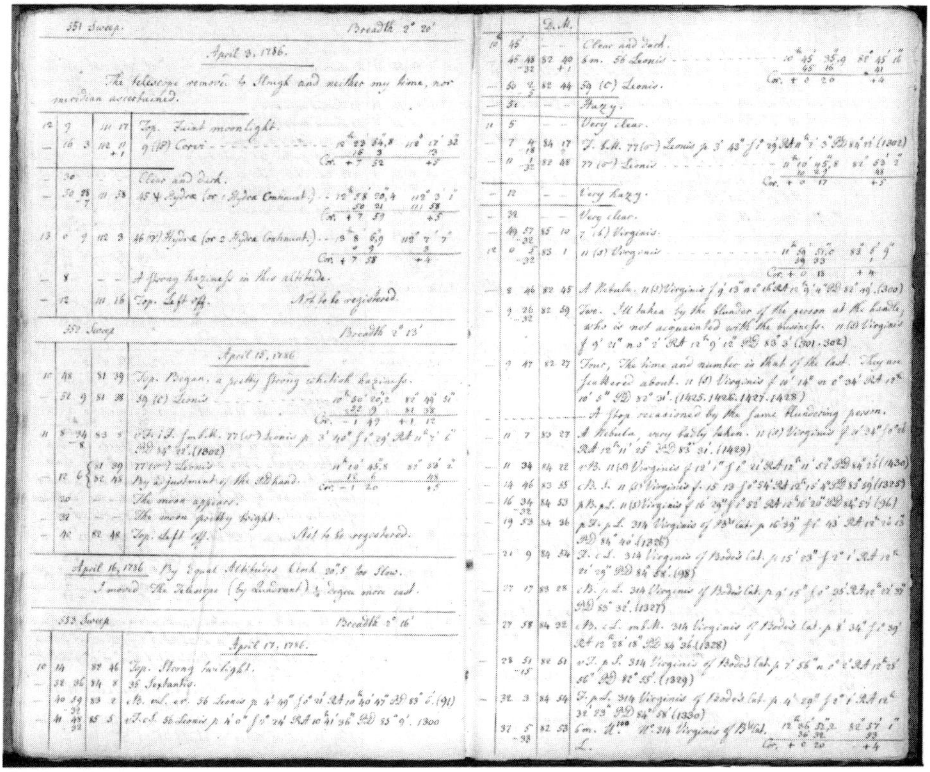

Figure 16. Caroline's fair copy of the first attempted sweeps for nebulae at Slough. For sweeps 351 and 352 William had not yet accurately determined either the direction of due south or the correct setting of the clock. As a result the observations were inexact in both coordinates and so were "Not to be registered." He made the necessary checks on April 16, and so sweep 553 on April 17 was valid. The number assigned to the first nebula discovered that evening, 91, is the number Caroline later gave it in the unified catalogue she prepared for William's son John after William's death. RAS W.2/3.5, courtesy of the Royal Astronomical Society.

involved. Only weeks after making the grant, he was demanding tangible signs of progress; and yet in the summer of 1786 he ordered William to go in person to Göttingen with the 10-foot he was giving them, oblivious to the huge disruption this would cause to the construction of the 40-foot. In William's absence Caroline coped as best she could, but she was more adept at receiving orders than giving them. The gardener was required for only three days a week, because in all the upheaval there was little he could

Figure 17. "An astronomer!!" An aquatint by the celebrated caricaturist Thomas Rowlandson (1756–1827). This is the first of Rowlandson's *Horse Accomplishments in 12 Sketches* (London, 1799). In each it is the horse rather than the rider that merits the title (in this case, by observing the heavens). Although there is no suggestion that this is a portrayal of William, the signpost pointing to the village of Slough ("SLOUHG"), where he lived, indicates that in popular imagination William was synonymous with "astronomer." Courtesy of Robert Anderson, photograph by permission of the Trustees of the National Museums of Scotland.

do beyond keeping the grass under control, "but he would be idling about the premises and gave me the name of stingy—in the Vilage; because I objected to his being there when not wanted."

On William's return construction got under way in earnest, as Caroline describes:

> The Garden and workrooms were swarming with labourers and workmen. Smiths and Carpenters going to and fro between the forge and 40 feet machinery and I ought not to forget; that there is not

one Screw-bolt about the whole apparatus but what was fixed under the immediate eye of my Brother. I have seen him lay stretched many an hour in a burning Sun, across the top beam when the iron work for the various motions was fixed.

At one time no less than 24 men (12 and 12 relieving one another) kept polishing day and night, and my Brother of course never leaving them all the while taking his food without allowing himself time to sit down to table.

The moonlight nights [when it was not possible to observe] generally were taken for such like experiments, and for the frequent Journeys to Town which he was obliged to make for ordering and providing Tools, and matherials which were continually wanted (I may say by whole sale).

William was in seventh heaven. His friend the novelist Fanny Burney records that the king's support for the great reflector "seems to have made him happier even than the pension, as it enables him to put in execution all his wonderful projects, from which his expectations of future discoveries are so sanguine as to make his present existence a state of almost perfect enjoyment. . . . He seems a man without a wish that has its object in the terrestrial globe."[14]

The mirror, which weighed nearly half a ton, presented a serious challenge. At the heart of William's success as an observer had been his genius in using his bare hands to impart to mirrors a shape that was close to the ideal parabola. But William needed a team of men to help grind and polish such a massive mirror, and they unhelpfully passed the time by "talking and sometimes singing on all sorts of subjects," the nature of which we can only guess. The result of course was that "those indications of the state of the mirror and of the polisher which were obtained by the touch when I polished by hand were intirely lost; so that my former experience became almost useless."

A visitor from Geneva has left us an account of the scene when the polishing was in its final stages:

In the middle of his workshop there rises a sort of altar; a massive structure terminating in a convex surface on which the mirror to be polished is to rest and to be figured by rubbing. To do this the mirror is encased in a sort of twelve-sided frame, out of which protrude as many handles which are held by twelve men. These sides are numbered and the men who are stationed at them carry the same numbers on the strong linen overalls which protect their clothes from the splashes of the liquid, which from time to time is introduced between the mirror and the mould to give the polish. The mirror is moved slowly on the mould, for several hours at a time and in certain directions. . . . It is then removed on a truck and carried to the tube, into which it is lowered by a machine expressly contrived for the purpose. This labour is repeated every day for a considerable time and by the observations he makes at night, Herschel judges how nearly the mirror is approaching the standard he desires.[15]

The mirror was first tried in February 1787, but it was too thin in the center and did not keep its shape when tilted in the tube. There was nothing for it but to order a much thicker mirror, weighing twice as much and therefore costing twice the sum budgeted. Clearly the grant of £2,000 was going to be nowhere near enough to pay for the reflector—and for its running costs, if and when it ever came into service.

In retrospect William can hardly be blamed for his miscalculation, for nothing on this scale had ever been attempted before. But one has a nagging suspicion that he may have been the first (but by no means the last) astronomer to make an unrealistically modest request to a funding body in the confident expectation that this body will eventually fork out whatever additional funds prove necessary, rather than see the original grant go down the drain. After all, the king's pocket was deep.

A Right Royal Row

It was at this point that William's normal good sense deserted him. He was of course unaware that the king's health was soon to show the first, tragic signs of the "madness" that would eventually lead to the appointment of his son as Prince Regent (in November the following year, the king was to tell an equerry he had seen Hanover through William's telescope[16]); but he must have known that the king would react badly if he had the slightest suspicion that he was being taken for granted. The king believed he had already fully funded the great telescope, and we can hardly blame him for becoming furious when he discovered that this was not the case. A row was inevitable.

The Herschels regularly dealt with disagreeable happenings by destroying any records and wiping the incident from their minds. For example, in 1792 their eldest brother, Jacob, was found strangled in a field outside Hanover, but of this trauma not the slightest hint survives in their voluminous archives, merely a letter from Dietrich to William thanking him for giving Dietrich his share in Jacob's estate; and Jacob is never ever mentioned again. Even so, the angry confrontation between William and the king left sufficient traces to shock William's son John when he came across the surviving evidence decades later. He sealed the documents in a bundle that was not to be opened until fifty years after his death, an instruction that his family meticulously honored. With these papers, and certain other clues that escaped destruction, we can piece together the story.

Understandably caught up in the excitement of building a telescope the like of which the world had never before seen, William considered the king's continued financial support to be the least of his worries. He simply wrote on July 18, 1787, to a royal aide enquiring whether he should ask the king for another block grant or merely send the bills for payment as they came in:

> As it was impossible to say exactly what sum might be sufficient to finish so grand a work, I now find that many of the parts take up much more time and labour of workmen, and more materials than I apprehended they would have taken, and that consequently my first estimate of the total expense will fall considerably short of the real amount. . . . I beg of you therefore to ask the King, whether it will please his Majesty that I should communicate the particulars of the further expense of the telescope to the President of the Royal Society, in order that he may, as before, take an opportunity to lay the same before the King, or whether his Majesty would order me to continue the workmen and apply from time to time for such sums as may be wanted.[17]

William saw the king as his enthusiastic ally in the great enterprise. It must have been about now that the Herschels gave a dinner party for musical friends including some of Sophia's sons. Outside, the great iron tube lay on the ground, nearly five feet in diameter. After dinner, no doubt inspired by the wine, the company assembled in the tube, where "God save the King" was sung to oboe accompaniment.[18] Not surprisingly, the diminutive Caroline "was one of the nimblest and foremost to get in and out of the tube."

The sheer scale of its aperture made the tube irresistible to visitors. On another occasion, Fanny Burney and friends walked through the tube, "and it held me quite upright, and without the least inconvenience; so would it have done had I been dressed in feathers and a bell hoop—such is its circumference." The Bishop of Worcester and a Dr. Douglas were watching, and when the others had gone, they too were tempted to make the same promenade.[19]

Unhappy at William's further demands for a project he believed he had already fully funded, the king decided the time had come to see for himself how things stood. A month later he descended in person on the Herschel home accompanied by a vast retinue that included the Archbishop of Canterbury. The king was in a good humor. Seeing the tube, he started to go inside it. The archbishop, being a foot or more taller than Caroline, hesitated to follow, at which the king said to him, "Come, my Lord Bishop, I will show you the way to Heaven." A few days later a euphoric William wrote to Sir Joseph Banks:

> as it is his Majesty's intention further to support the construction and completion of this instrument as well as to provide for such necessary annual expences as will be connected with its being kept up and serving for a series of observations, I shall lay before you an account of the things which are still wanting with an estimate of the expences they may occasion.

These came to nearly £1,000 for the construction, and running costs that he would try to keep down to £200 a year.

William then came to a delicate further matter: the running costs of Caroline. Choosing his words carefully, William explained that Caroline was more than happy to continue as his assistant, but that she could not continue doing this for love because she needed financial security. Playing the gender card, he enquired whether perhaps the queen would allow her an annual pension of £50? The alternative would be for William to hire a replacement at twice the cost.

Banks's response on behalf of the king was ominously curt:

> I have this moment seen the King who has granted all you ask but upon certain conditions which I must explain to you. Will you be

so good as to come to me in Soho Square tomorrow as soon as convenient that we may finish this matter & that I may report to him
before he sets out for Windsor.

Forty years later, Caroline was to write to John with the merest hint of the
row that ensued.

> But there can be no harm in telling my own dear nephew, that I
> never felt satisfied with the support your father received towards his
> undertakings, and far less with the ungracious manner in which it
> was granted. For the last sum came with a message that more must
> never be asked for. (Oh! how degraded I felt even for myself when-
> ever I thought of it!)

Caroline writes as though the dealings between the king and the court astronomer were by correspondence only, but it is evident that in fact there
was a face-to-face confrontation between them that left William outraged
and humiliated. He shared his troubles with Watson, who had done so
much to promote William's cause at court. This letter is lost, but Watson's
extraordinary attempt at consolation survives:

> I do most sincerely sympathize with you, & feel in some measure as
> you must feel at the unworthy treatment you (& I may add Science)
> has received. But I sincerely hope by the latter part of your letter that
> the Storm is past, & that the K— is brought by reflexion to know
> you . . . a little better.

After three pages in this vein he concludes: "You must send me a letter in
your old style of writing to efface from my mind the mortifying traces of
your last."

The most remarkable aspect of this "unworthy treatment" is that the
king in fact gave William far more money than he asked for. One might
have supposed that the prudent Banks had ensured there would be no
third application by adding a sum for contingencies before passing William's application to the king, but the letter in the royal archives matches
William's file copy exactly. We must conclude that the king told William
in no uncertain terms that he was shocked to find the original grant had
not been enough to built the reflector, as he had been promised it would;
that he would reluctantly approve the present request and, since he had no
confidence that William had got his sums right this time, he would add a

further margin for error; but that this was absolutely the last subvention he would provide, and furthermore, that he expected every penny of it to be accounted for.

It was. Twelve men engaged in polishing for six weeks were expected to drink a pint of beer each day (surely an underestimate) and this would total £3 12s. In winter time, "Two fires and four or five candles all night, when the weather is fine. One fire day and night the whole winter. In the day for company that comes to see the telescope, in the night waiting the coming of the stars tho' the weather should be cloudy"—William is required to reckon the cost of every candle and of every log of wood for the fire.

Everyone knew that the king had given a second grant similar in scale to the first, and it suited both the king and William to maintain in public the pretence that this grant had been awarded as cordially and willingly as the first. But the private relations between king and court astronomer never recovered, and one suspects that the fracas was the reason why William's knighthood had to wait until 1816 and the regency of George III's son. The honor, when it came, was long, long overdue. Caroline later wrote to William's son John, "General Komarzewsky used to say to your father, Why does not he (meaning King George III) make you Duke of Slough?"[20]

The second and thicker mirror was successfully cast in February 1788. As many as twenty-two men were required for the polishing, depending upon the stroke being used. In October 1788 it was tried on Saturn, but the focus was too long. Three months later William began to experiment with a machine to polish a 20-foot mirror, and after two months this was proving sufficiently successful for him to make a first attempt to machine-polish the thin mirror for the 40-foot. Alexander contributed his practical ingenuity by making sketches of how such machines might work. William then made a smaller machine, for a 7-foot mirror, and in June returned to the thin 40-foot, but "only by way of trying the apparatus." At last, in July and August, he was able to work on the thick 40-foot mirror.

Meanwhile construction of the ladder-work mounting to support and direct the great iron tube was going ahead. It was a scaled-up version of the mounting of the large 20-foot, but with many elaborations. No longer would Caroline be seated at a nearby window. To one side of the base of the tube was a hut for her and the gadgets she needed, linked to William by a speaking tube, while to the other side was a small hut for the work-men. Two men were needed to raise or lower the observing platform, and they must work in unison if the platform was not to tilt disastrously. It is probable that it was Alexander who devised the bell mechanisms, one for each man; the mechanisms would ring in unison if the maneuver was proceeding correctly.

The astronomical community awaited with bated breath the completion of the great reflector, with expectations not unlike those that would one day greet the launch of the Hubble Space Telescope. Jérôme de Lalande, for example, had written from the Collège Royale in Paris in November 1786 to say that he planned next summer to visit England "voir votre superbe telescope de 40 piedes." The following May he asked whether the instrument was now operational, for if so he would make the journey; and a year later he wrote again in the same vein. Banks, Watson, and the British scientific community had matching expectations, for patronage of science on this scale was unheard of. William was under huge pressure to deliver.

Industry Rewarded: Saturn's Sixth and Seventh Moons

Saturn came to his rescue. It had been the very first object he had "officially" observed back in March 1774 when he opened his first observing journal. As it happened, Saturn's ring was then almost edge-on to the terrestrial observer. When this is the case, the glare of the ring is reduced almost to zero; and if any of the planet's moons are located in the plane of the ring, as is more than likely, they may well become visible, strung out in a straight line. As the 40-foot neared completion, Saturn was nearly edge-on once more, and when Lalande wrote to William in November 1786, he urged him to use the great reflector to study the planet.

In August and September 1787, with the 40-foot still unfinished, William was sufficiently fascinated by Lalande's suggestion to examine Saturn's moons with the 20-foot on no fewer than seventeen nights. He was intrigued with the challenge of making sense of their complex ballet around their parent planet. On August 19 he observed them for three hours on end and thought he had discovered a sixth moon; but observations in the weeks that followed failed to produce proof of this, partly because of the difficulty William had in distinguishing moons from background stars.

Next summer he again gave Saturn time on the 20-foot, observing on one occasion for no less than five hours. But in its regular work the reflector was currently sweeping for nebulae near the zenith, and while this was in progress, its mirror was horizontal. As a result dew collected on it, and the mirror became tarnished and so was less suited for the delicate observations involved in tracking a planet's moons.

By the summer of the following year, 1789, conditions were almost ideal. The second mirror of the 40-foot was nearing completion; Saturn's ring was almost exactly edge-on; and the planet was now moving rapidly backward across the sky, taking its moons with it but leaving background stars behind. And Lalande had just sent him new tables of the motions of the five known moons. William observed the planet for six nights with the 20-foot, and then, on August 28, 1789, he directed the 40-foot toward it (figure 18). Only four of the five known moons were then visible, but he could see what looked like a sixth. "What makes me take it immediately for a satellite," he noted, "is its exactly ranging with the other four and the ring."

William was under great pressure to justify the money and effort that had gone into the 40-foot, and the moon of Saturn was the answer to his prayers. August 28, he declared, marked the commissioning of the great reflector, and a memorable night it had been. The master builder had pointed the monster toward its first target in the sky, and, hey presto!, Saturn had a sixth moon. It was nothing short of magic. William dashed off a letter to Banks asking him to add a brief announcement of the discovery to a paper already in press: "P.S. Saturn has six satellites. 40 feet reflector."

In this letter, William had been careful to do "my good 20 feet-Telescope the justice to say that strong suspicions of the 6ᵗʰ satellites existence were given by it." Indeed, it soon occurred to him that there might possibly be a priority dispute in the making: no doubt he was not the only observer whom Lalande had urged to search for Saturnian satellites, and perhaps he had a rival, as yet unknown, in the discovery of the sixth. He therefore wrote again to Banks, this time to say that "to my surprise" he now realized that he had in fact discovered the sixth satellite with the 20-foot back in August 1787; but that he had not followed this up because he was concentrating at the time on the moons of Uranus. As proof of the discovery he cited the exact time of the satellite's greatest western elongation: 1787 August 19ᵗʰ 21ʰ 54' 56" sidereal time.

The claim was bizarre, for if sight of an object without demonstrating its true nature constituted "discovery," then Flamsteed discovered Uranus in 1714 and Galileo discovered Neptune in 1612.[21] But William now had

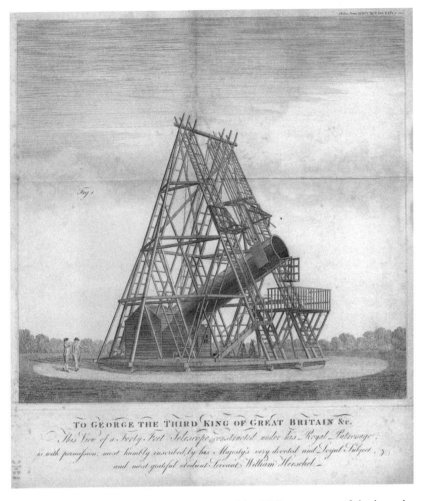

Figure 18. The 40-foot reflector, as illustrated by William as part of the long description of the instrument he published in *Philosophical Transactions* in 1795.

reason in his postscript to omit mention of the 40-foot, and so he proposed to Banks a modified wording: "Saturn has six satellites. An account of its discovery, its Revolution & orbit will be given in the next volume of the Ph[ilosophical] Tr[ansactions]." If priority became an issue, he would argue that he had discovered the satellite in 1787 with the 20-foot; if not, the discovery would be credited to the 40-foot on the night in 1789 when it first came into service.

Banks got the message: "Your two letters I shall keep as they are authentic Papers relative to the dates of your discovery which may hereafter, tho' I do not think they will, be called for." But it soon became clear that there were no rival claims to the sixth satellite, and in the postscript as published, the discovery was credited to the 40-foot.

Things then got even better. On September 8, when viewing Saturn with the 20-foot, William suspected there was a seventh moon. He confirmed this with the 40-foot on the seventeenth: "I see six satellites at once, and being perfectly assured that the 2^d is invisible it becomes evident that Saturn has 7 satellites." Euphoric, he wrote again to Banks: "Perhaps I ought to make an apology for troubling you with a letter on the same subject as my former one; but if satellites will come in the way of my 40 feet Reflector, it is a little hard to resist discovering them."

Banks had invested heavily in William and his giant reflector. He had persistently lobbied the king on William's behalf in the aftermath of the discovery of Uranus, and he had sponsored both applications for funding for the 40-foot (on the second occasion being snubbed for his pains)—and so, like William, he badly needed the 40-foot to work miracles. He too was euphoric at the news from Slough, so much so that his usual good sense deserted him and he looked forward to an endless succession of new discoveries: "We expect daily to partake of new wonders from the effects of the 40 feet you may if you please say we are insatiable not to be content with 2 satellites, in truth if nothing else appears we shall be happy in the possession of them and gratefull to the discoverer but you will not wonder if our hopes are at the period a little elevated."[22]

The most optimistic expectations of the king, the Royal Society, and the European astronomical community at large were being fulfilled, and with astonishing speed. In truth William had no hope whatever of sustaining this level of performance; but for the moment he allowed himself to bask in the glow of triumph. In his report in *Philosophical Transactions* William paid a fulsome tribute to his royal patron. Knowing what we do of their clash, we may detect a hint of "Oh ye of little faith!":

> But it will be seen presently, from the situation and size of the satellites, that we could hardly expect to discover them till a telescope of the dimensions and aperture of my forty-feet reflector should be constructed; and I need not observe how much we Members of this Society must feel ourselves obliged to our Royal Patron, for his

encouragement of the sciences, when we perceive that the discovery of these satellites is intirely owing to the liberal support whereby our most benevolent King has enabled his humble astronomer to complete the arduous undertaking of constructing this instrument.

The two men—William and King George—who in their different ways had come together to create the monster, had had quite different agendas. The king was promoting his favorite science, astronomy (and gaining plaudits for so doing), and he was incidentally solving the problem of how to entertain guests after dinner at the castle. To William, the huge mirror was intended above all to collect enough light from very distant nebulae for him to examine them and confirm that they were simply clusters of stars.

True Nebulosity Once More

Then it all went wrong. At 3:51 a.m. on November 13, 1790, William was at the eyepiece of the 20-foot engaged in a routine sweep, number 980 in the series, with Caroline at her desk at a nearby window. There was no reason to expect that something dramatic was about to happen. But, as the sky slowly rotated overhead, a bizarre and ghostly object came into his field of view: a star with an atmosphere (Plate 5). Caroline copied down the description at William's shouted dictation:

> A most singular phaenomenon! A star of about the 8th magnitude, with a faint luminous atmosphere, of a circular form, and of about 3' in diameter. The star is perfectly in the center, and the atmosphere is so diluted, faint, and equal throughout, that there can be no surmise of its consisting of stars; nor can there be a doubt of the evident connection between the atmosphere and the star.

Until now, the "planetary nebulae" that William had come across had been too far away for him to make out the central star that we know to have been there. To him, they had the circular, disklike shape of a planet, and the faint light of a nebula, and this was why he had coined the name we still use today. But the "singular phaenomenon" was a planetary nebula so near that the central star was obvious to him at a glance, as was the cloud (of gas) surrounding it.

This was not the first example of a star apparently associated with nebulosity that he had come across in recent years, but it was the first that he had no hope of explaining away on his current theory that all nebulae were star clusters. Here indeed was "true nebulosity." And in a brilliant stroke, William took his theory of evolution under the action of gravity back to the time before the star was fully formed. The luminous matter was "fit to produce a star by its condensation"—in other words, gravity was working on the luminous matter to bring about the birth of the star. But the question that had led him to build his monster telescope had now been answered, and William was left with a cumbersome and unwieldy reflector that was of little use. Yet the king had spent a fortune on its construction, and royal guests would continue to trek to Slough to admire the fruit of his munificent patronage of science. And the astronomical world looked forward confidently to a series of unparalleled discoveries. William had made a rod for his own back.

8

1786–1788
"Gold Can Glitter as Well as the Stars"

Caroline Discovers Her First Comet

When William and Caroline moved to the Grove in April 1786, they found themselves living a stone's throw from their landlady, Elizabeth, widow of Adee Baldwin. Adee was the grandson of William Baldwin, who in 1701 had been granted a long lease on the Crown Inn that stood on the corner of Windsor Road and the Great West Road, and the son of Thomas Baldwin, who had instituted the first daily coach between London and Bath. The coach had evidently proved a gold mine. At the time of his death Adee not only leased the Crown Inn but also owned several other Slough properties, including the land and dwellings to the south of the inn (figure 19) as far as the Herschel home. These had now passed to his widow.

Adee and Elizabeth had two surviving children, Mary (plate 6) and Thomas. Mary had herself married into money, but money does not buy health, and her husband John Pitt, a "gentleman" who lived off the rents of properties close to Slough that included an inn on the Great West Road, was ailing. He and Mary lived close by in the next village, Upton, and their land and William's shared a common boundary. Many an evening, when the weather was poor and observing impossible, William and Caroline would stroll over to the Pitts' home (figure 20) and spend the evening with them.

But before long, King George was insisting that William go to Göttingen in person to present one of the 10-foot reflectors he had commissioned. Fortunately it was when Bath was out of season, and so Alexander was able to accompany him. They left at the beginning of July 1786 and returned in the middle of August, having visited their family in Hanover.

In their absence Alexander's wife, Margaret, came to stay with Caroline. One wonders with what success Caroline was able to hide her impatience

Figure 19. Windsor Road, Slough, from a photograph in Herschel Family Archives. The large building on the left is the Crown Inn, on the southeast corner of the crossing of Windsor Road and the Great West Road from London to Bath. Next to it is the cottage occupied by the widowed Elizabeth Baldwin. She was leaseholder of the inn and owner of several properties in the area, including the Herschel home (which was on the Windsor Road, some 200 yards from the inn).

at her sister-in-law, for she writes, "I was obliged frequently to sacrifice an hour to her gossipings." Caroline was doing her best to keep work on the construction of the 40-foot moving forward, but she felt she lacked the proper authority to give orders. There was, beside, a mountain of deskwork for her in preparing the latest catalogue of nebulae for publication. But for the time being she was free from her nighttime duties as amanuensis, and so at last she had the opportunity to observe on her own account.

With exciting results. August 1: "I have calculated 100 nebulae today, and this evening I saw an object which I believe will prove tomorrow night to be a comet." August 2: "Today I calculated 150 nebulae. I fear it will not be clear to-night, it has been raining throughout the whole day, but seems now to clear up a little. 1 o'clock; the object of last night *is a Comet*. I did not go to rest till I had wrote to Dr. Blagden and Mr. Aubert to announce the comet."

Figure 20. (*above*) Upton House, c. 1700; (*below*) the same house (now 74 Upton Road, Slough) today. When William moved to Slough, Upton House was owned by Mrs. Elizabeth Baldwin and occupied by Mrs. Baldwin's daughter Mary and Mary's husband, John Pitt. After the death of John Pitt and Mary's subsequent marriage to William Herschel, Mary and William for a time maintained their two establishments as equal homes. Mary later leased Upton House to tenants, and from 1803 to 1810 Caroline rented rooms from the tenants. Upton House is now the only Herschel home in the Windsor area to survive. Photograph courtesy of Michael Day and www.SloughHistoryOnline.org.uk.

Alexander Aubert was one of the Herschels' most loyal allies, and his letter of congratulations to Caroline was fulsome indeed: "You have immortalized your name and you deserve such a reward from the Being who has ordered all things to move as we find them." Charles Blagden was secretary of the Royal Society, and Caroline's letter reached him just before

the annual visitation of the Royal Observatory at Greenwich, when leading astronomers heard a report from the Astronomer Royal, whose work
they nominally supervised. He replied to Caroline: "most of the principal
astronomers in and near London attended, which afforded an opportunity
of spreading the news of your discovery, and I doubt not but many of them
will verify it the next clear night." Next Sunday, he continues, Sir Joseph
Banks and friends "may wait upon you to beg the favour of viewing this
phenomenon through your telescope." And so it was that on August 6,
little Caroline, the scullery maid from Hanover, entertained the president
and secretary of the Royal Society, along with Lord Palmerston, all men
of the London establishment who had journeyed to Slough for the sole
purpose of seeing Caroline's comet through Caroline's telescope.

Caroline also wrote to her brother Dietrich in Hanover to tell him
about her comet and how he could find it with the telescope William had
made for him when they were in Bath together. She told him how pleased
she was to learn from William that he and Alexander found Anna in good
health. "I hope our dear mother does not grieve too much now they have
left her. I dare say William will pay soon another visit, and then I will take
that opportunity of coming to see her." In middle life, Caroline retained
her affection for Anna and showed no resentment for the harsh treatment
she had endured at her hands. But in old age bitterness would enter Caroline's soul, and then she would exact a terrible revenge.

William arrived home a week later to learn that his little sister had
become famous in his absence and to find a summons to Windsor Castle
where he was to demonstrate Caroline's comet to the royal family. Fanny
Burney was present. "The comet was very small, and had nothing grand or
striking in its appearance; but it is the first lady's comet, and I was very desirous to see it." She also painted a charming picture of its discoverer: "She
is very little, very gentle, very modest, very ingenuous; and her manners
are those of a person unhackneyed and unawed by the world, yet desirous
to meet and return its smiles."[1]

William Courts and Marries Widow Pitt

But there was a cloud on Caroline's horizon. During the summer of 1786
John Pitt died, and to Caroline's dismay, William (now "Dr. Herschel"
of Edinburgh University) began to court the "sensible, good-humoured,

unpretending and harmless" widow.[2] As Fanny Burney wrote, "she was rich too! and astronomers are as able as other men to discern that gold can glitter as well as the stars."[3]

Mary was rich and would one day be much richer. John Pitt had left her well provided for,[4] but his legacy paled into insignificance compared to what she might expect on the death of her widowed mother. Her grandmother on her father's side, Grace Baldwin, had come from a large family, and one of Grace's sisters had married into money. The sister's son, a lawyer named Nathaniel Phillips, was childless, and he left his estates in Wiltshire not to his own two sisters but to his cousin Adee Baldwin, Grace's son and Mary's father. It seems the sisters put up a fight, for the transfer had to be confirmed by decree of the High Court of Chancery; and well they might, as the estates were worth some £30,000.[5]

Adee Baldwin was now wealthy beyond the dreams of avarice. Not only did he have the lease of the Crown Inn and his flourishing coaching business, but he had estates worth 150 times William Herschel's annual pension from the Crown. And he had himself married into money. We know this because his wife, Elizabeth, had a sister Mary, known in the family as "Aunt Clark." Mary survived a tempestuous marriage that ended with her being virtually excluded from her husband's will,[6] but nevertheless Mary ended her life in possession of well in excess of £12,000.[7] This money could only have come to her by inheritance, and Elizabeth Baldwin must have inherited a similar sum. Mary Pitt would one day share all this wealth with her brother Thomas.

Yet, in spite of this, Mary was bored. A neighbor, Charlotte Papendiek, records that the thirty-six-year-old Mary,

> poor woman, complained much of the dullness of her life, and we did our best to cheer her, as also did Dr Herschel, who often walked over to her house with his sister of an evening, and as often induced her to join his snug dinner at Slough.

Mary's surviving son by her first marriage, Paul Adee Pitt, was then aged twelve or thirteen, and doubtless away at school.

> Among friends it was soon discovered that an earthly star attracted the attention of Dr Herschel. An offer was made to Widow Pitt, and accepted. They were to live at Upton, and Miss Herschel at Slough, which would remain the house of business.

Mary then had second thoughts. William would be all day at Slough, where the 40-foot was under construction, and half the night as well, sweeping for nebulae. He would spend more time with his sister than with his wife. So she called off the engagement.

William, to his credit, would not abandon either astronomy or Caroline, although Mary's money would make his salary from the king an irrelevance:

> Dr Herschel expressed his disappointment, but said that his pursuit he would not relinquish; that he must have a constant assistant and that he had trained his sister to be a most efficient one. She was indefatigable, and from her affection for him would make any sacrifice to promote his happiness.

Months passed, and in the autumn of 1787 a bizarre compromise was agreed. They were to maintain the two houses on an equal footing, with accommodation in each, two maidservants in each, and a footman to go between the two.

William, in a still more extraordinary move, then asked Watson to take soundings among astronomers as to whether they approved of his marrying—for the stars would now have to compete with Mary for his attentions at night. What, we might ask ourselves, did he intend to do if they did not approve? Fortunately Watson was able to report that "excepting some little fears with respect to Astronomy," William's plans "seemed to meet with the approbation of everyone." He personally thought William might even contribute more to astronomy if he adopted a more relaxed lifestyle.

The consent of the astronomical community having been given, the wedding went ahead. On May 8, 1788, William and Mary were married in the little church of Upton, whose parish included Slough. Sir Joseph Banks came down from London to be best man, and Alexander journeyed from Bath to be a witness along with Caroline. Mary was given away by her brother Thomas, a wholesale chemist or "druggist." Paul, who was going into business with Thomas, was also present. The reception took place in the newlyweds' Slough home rather than in Mary's house in Upton, and it was in Slough that they spent their honeymoon. However, it was from Upton that the customary cards were sent to friends a few weeks after the wedding, and so it was there that the friends called to offer their congratulations.

The dual establishments survived for a while, sometimes to the bemusement of their friends. Charlotte Papendiek tells us of the one and only occasion on which she and her husband were invited to dine with William and Mary.[8] They chose to travel independently of each other, and Charlotte went by stagecoach. "I had carelessly read the note of invitation, and knowing that the Doctor always did remain at Slough during the winter, so as to be on the spot for his observations, I took it for granted that they were there now, so took the stage to that place." When she discovered her mistake, the kindly coachman offered to drop her off at a place where she would have only one field to cross in order to reach Mary's home. However, it turned out that the field contained not only cows but a bull, "and down I fell from terror and the damp, slippery ground." The Herschels were no doubt astonished at the bedraggled spectacle that Charlotte presented when at last she reached the house. She was given a room in which to repair her appearance, "and bathed my hand and arm, which were much swollen from the fall, and in great pain. A glass of wine revived me, and the dinner went off well." The arrangements for her return home were better, but not much. "Thus shabbily ended this invitation, which the Herschels did not repeat."

Paul was still writing to his mother "at Upton" as late as May 1789, one year after her wedding, but by September 1791 Mary—then pregnant—had accepted that it made more sense to join William in Slough and to rent out her Upton house.

Back in the summer of 1787, in expectation of marrying into money, William had felt able to offer financial provision for Caroline, who had hitherto worked for bed and board; but his well-intentioned offer was far from welcome. The outraged Caroline had had enough of brotherly handouts. For sixteen years she had dedicated herself to William and had given up her own career as a singer so that she could be of service to him. She had run his home, copied his music, trained his choirs, spent nights in the depths of winter taking down his shouted observations when she could have been in bed, transcribed hundreds of pages of symbols at his behest. Though she had been "the keeper of my brother's purse, with a charge to provide for my personal wants, only annexing in my accounts

the memorandum <u>for Car.</u> to the sums so laid out—when cast up, they
hardly amounted to seven or eight pounds per year." Now another woman
had primacy in his affections and at his table. She demanded that William
ask the king for her to have a proper salary as his paid assistant; and, as
we have seen, his request was granted. The initial installment was "the first
money I ever in all my lifetime thought myself to be at liberty to spend
to my own liking," although extracting her "pension" from the royal purse
was to prove no easy matter.

The pension was consolation but little comfort. Caroline's nose had
been put severely out of joint by William's marriage. But slowly Mary won
her round. As Caroline's great-niece would one day write, "Miss Herschel's
good sense soon got over the startling innovation of an English lady-wife
taking possession of her own peculiar fortress, and she who gladdened her
husband's home soon won the entire affection of the tough little German
sister."[9]

Looking back in later life, Caroline was ashamed to read what she had
written in her diaries for the years immediately after William's marriage. So
she destroyed them, to the immense vexation of later historians.

Curiously, in 1803 Caroline agreed to accept from Mary £10 a quarter, and thereafter "Miss H—, £10" appears regularly in Mary's accounts.
Caroline was fiercely independent and resentful of her lifelong depen-
dence on brotherly handouts, and her acceptance of the quarterly allowance is surprising. She herself felt uneasy about it, for after her return to
Hanover, in 1824, she felt it necessary in a letter to Mary—now a valued
friend—to claim that "I certainly should not have accepted it if I had not
been in a panic for my friends in Hanover," namely her brother Dietrich
and his family.

Mary Herschel's late husband John had left her and their son, Paul,
well-provided for—£2,000 cash in trust for Paul, and for Mary a life interest in John's estate. In 1793 Paul died when only nineteen, and his £2,000
came to Mary. In 1795 Mary's Aunt Clark died, leaving Mary and her son
John jointly £5,000 of 3 percent stock, and to Mary and her brother a half-
share each in her house at Walton, the Crown Inn in the same village, and
property in Greenwich.[10] In 1798 Mary's mother followed Aunt Clark to
the grave, leaving Mary the house and property where Mary and William
lived, the adjacent garden, and a half-share in the rest of her vast estate.[11]

If Mary had a financial problem, it was in keeping track of all her wealth. William's pension of £200 a year was no longer of any consequence— although this did not stop him claiming it, just as he drew the allowance for staffing the 40-foot, even when no such staff had been employed.

9

1788–1798
"Noble and Worthy Priestess of the New Heavens"

Caroline the Comet Hunter

A fortnight after William's wedding on May 8, 1788, he and Caroline resumed their nighttime sweeps for nebulae. These were now less frequent, for reasons we can all understand, and so Caroline's duties as amanuensis were much reduced. In the daytime the Herschel home was a building site, with the construction of the 40-foot in full swing, and in this Caroline had no part. And she was no longer mistress of his household, with responsibility for managing the servants and keeping track of the expenditure. Instead, she was a spinster living on her own in the adjoining cottage, with a flat roof on which to position her observing instruments, and time on her hands during which she could sweep for comets.[1] It seems that she had sole use of the clock that Alexander had made for William, and he made another to meet her need for a clock to have by her when she was observing. Known in the family as "the monkey-clock"—perhaps a play on the name of John Monk, principal maker of such clocks at the time—it was "to take with me on the roof when I was sweeping for comets, that I might count seconds by it going softly downstairs till I was within hearing of the beat of the timepiece on the first floor." In this way Caroline could determine the exact time of the observation she had just made.

The Astronomer Royal tells us how his "sister astronomer" worked in a letter he wrote in 1793, by which time William had made her a second and larger sweeper:

> I paid Dʳ & Miss Herschel a visit 7 weeks ago. She shewed me her 5 feet Newtonian telescope made for her by her brother for sweeping the heavens. It has an aperture of 9 inches, but magnifies only

from 25 to 30 times, & takes in a field of 1°49' being designed to shew objects very bright, for the better discovering any new visitor to our system, that is Comets, or any undiscovered nebulae. It is a very powerful instrument, & shews objects very well. It is mounted upon an upright axis, or spindle, and turns round by only pushing or pulling the telescope; it is moved easily in altitude by strings in the manner Newtonian telescopes have been used formerly. The height of the eye-glass is altered but little in sweeping from the horizon to the zenith. This she does and down again in 6 or 8 minutes, & then moves the telescope a little forward in azimuth, & sweeps another portion of the heavens in like manner. She will thus sweep a quarter of the heavens in one night. The D^r has given her written instructions how to proceed, and she knows all the nebulae [listed by Messier] at sight, which he esteems necessary to distinguish new Comets that may appear from them. Thus you see, wherever she sweeps in fine weather nothing can escape her.[2]

But he thought she would do well to follow the Greenwich practice when searching for comets and briefly look over the whole sky with a simple night glass before settling down to more detailed observations; and in 1800 he presented her with a field glass and a pair of binoculars to encourage her to do just this.

In fact Caroline's searches were even more versatile than Maskelyne implies. In the middle of the night, when the Sun was long gone, Caroline would indeed sweep large areas of sky in a vertical direction, from the horizon to the zenith and back. But because comets swing around the Sun in their passage through the solar system, the sky in the west just after sunset and in the east before sunrise are prime hunting grounds for the comet seeker, and at such times Caroline would search with a horizontal movement.

But comets are rare, and almost invariably her sweeps yielded nothing. She wearied of writing endless negative entries in her observing book; but she became concerned that the book might one day fall into the hands of an unsympathetic reader, who would think she had done nothing to justify her salary from the Crown. On one occasion she wrote: "I have kept no memorandum of my sweepings, tho' I believe I may say that I have neglected no opportunities whenever they offered; but, not meeting with any comet, I looked upon keeping memorandums of disappointments as time thrown away."

Caroline's Second Comet

In the middle weeks of December 1788 she tried in vain to locate a comet that had recently been discovered by Messier and that William reckoned would now be near the North Pole. She then began to search for the predicted return of the great comet of 1661, which was (wrongly) suspected of having itself been a return of the comet of 1532 and if so might shortly reappear. She was looking for it on December 18, when William decided to sweep for nebulae with the 20-foot and she was required to assist him. On the nineteenth she swept on her own account, but the next night she had again to act as William's amanuensis. On the twenty-first she resumed where she had left off on the nineteenth, and "when I had swept as far as Beta Lyrae, I perceived a comet." She was still observing at 5:30 in the morning. This, the second of Caroline's comets, reappeared in 1939 and is expected again in 2092.

Nevil Maskelyne was particularly impressed by this discovery. "As it came up from the south it seems that Miss Herschel lost no time in finding it. I mean that it could not have been seen much sooner even in her excellent telescope." It was, he wrote to William, a "Cometic Ghost, conjured up by your sister, with her powerful glass instead of a wand."

Caroline's Third Comet

It took her just over a year to find her next comet, "an object with a burr all round," which she discovered on January 7, 1790. It was probably this success that prompted William, for whom bigger was always better, to construct for her a larger sweeper, with an aperture more than twice the diameter—and four times the surface area—of its predecessor. Unfortunately the focal length was more than Caroline measured from head to toe, and so to reach the eyepiece she needed to stand on a stool, and this she found tiresome. Sometimes she used one sweeper and sometimes the other, but the "small" sweeper had a special place in her affections, and in later life she was to take it with her to her native Hanover.

Caroline's Fourth Comet

In 1783, and again in 1787, William had seen bright spots on the Moon that he had interpreted as active volcanoes, and from time to time Caroline examined the Moon in hopes of finding further evidence of volcanic activity. She did so on April 16 and 17, 1790, and on the second evening she also found her fourth comet. As usual, she was only too happy to hand over her comet to the professionals; but this time there was an unfortunate delay.

Whereas the planets orbit the Sun in paths that are nearly circular, the orbits of comets are highly elongated. Pulled by the Sun's gravity, they appear from outer space, swing around the Sun, and soon disappear once more. We glimpse them only briefly, and so eighteenth-century astronomers faced a daunting task in trying to establish the shape of a comet's orbit: was it an elongated ellipse, in which case the comet would reappear at some future date, or was it parabolic or hyperbolic, in which case the comet would depart never to be seen again by human eyes? Every night of accurate observation was precious in deciding between the two, and Maskelyne at Greenwich was driven to distraction by the time it took for news of Caroline's discoveries to reach him (the penny post could take as much as two days!) and by the amateurish way in which she specified the comet's location before he could take it into his own care. In this particular instance, delays in the post and inclement weather cost him no fewer than six nights. He tried, as patiently as he could, to explain to Caroline how to define a comet's position, and he begged her to send him word of a new discovery as quickly as express post would allow—he would be happy to bear the cost.

Caroline's Fifth, Sixth, and Seventh Comets

Caroline found her fifth comet in 1791, her sixth in 1793 (but Messier had already noticed it), and her seventh in 1795. A quarter of a century later, the German astronomer Johann Franz Encke was to show that this same comet had been seen by Pierre Méchain back in 1786 and that, astonishingly, it returns every three years or so. We know it as Encke's Comet.

Caroline's discoveries won her a reputation both national (plate 7) and international, and male astronomers across Europe competed to flatter their one and only female colleague. Professor Karl Felix Seyffer of Göttingen was the outright winner: "Permit me, most revered lady, to bring to your remembrance a man who has held you in the highest esteem ever since he had the good fortune to enter the Temple of Urania, at Slough, and to pay his respects to its priestess. . . . Give me leave, noble and worthy priestess of the new heavens, to lay at your feet my small offering of eclipses of the sun, and at the same time to express my gratitude and deepest reverance."[3] And so on and so on.

Caroline's Eighth Comet

Caroline discovered her eighth comet with the naked eye, on August 14, 1797. As luck would have it, William was away, and so Caroline had to decide for herself how best to get word to Maskelyne. She adopted the simplest method: she would go in person. It was years since she had ridden on horseback further than the couple of miles to Windsor, and it was twenty miles to London and then another six or seven to Greenwich. But needs must. . . . She allowed herself an hour's nap, saddled a horse, and later that day presented herself to an astonished Maskelyne. She stayed with him a night or two and then set off on the return journey to Slough. Maskelyne urged her to call in on Sir Joseph Banks in London and give him the news, but her shyness got the better of her. "I thought a woman who knows so little of the world ought not to aim at such an honour, but go home, where she ought to be, as soon as possible."

Caroline's Ninth Comet?

Caroline laid claim to eight comets, and historians have agreed. But it seems she may well have discovered a ninth. Back in 1783, when nebulae were uppermost in her mind, she had noticed on July 30 that "In the neck of Equuleus or head of Aquarius there is a rich spot; near 3, 4, 7 Equulei"; and again, on August 24, "Between gamma & delta of Equulei a rich spot." These are the only two occasions on which she records ever seeing "a rich spot," and there are no plausible nebulae in the locations she specified. It seems likely that in fact Caroline twice glimpsed a comet that escaped all other observers (and so is otherwise unknown to astronomy), her rival

Plate 1. This charming miniature of William (artist unknown) is the only representation of him as a young man. Caroline in later life gave it to William's son John, and it is likely that William had given it to Caroline when he visited Hanover in 1764. Herschel Family Archives.

Plate 2. The "large" 20-foot reflector at Datchet, watercolor by Rev. Thomas Rackett (1755–1840), showing William in the observing chair (in broad daylight!). When the reflector was commissioned in October 1783, the actual mounting of the reflector was fixed facing south, and William "swept" by standing on an observing platform and dragging the tube from side to side, through an arc of 12 or 14 degrees. By the end of 1783, however, William was sweeping by using the reflector as a transit instrument with the tube itself facing due south while the heavens rotated overhead. He might then replace the observing platform with a simple chair, as shown here. In the painting, we see that the mounting can be rotated, a development introduced no later than September 1784; the painting therefore may be dated to the second half of 1784 or the first half of 1785, after which the Herschels moved from Datchet to Old Windsor. We are shown the workman raising and lowering the tube by means of a windlass; a second windlass, to the workman's left, allowed him to raise or lower the chair. Herschel Family Archives.

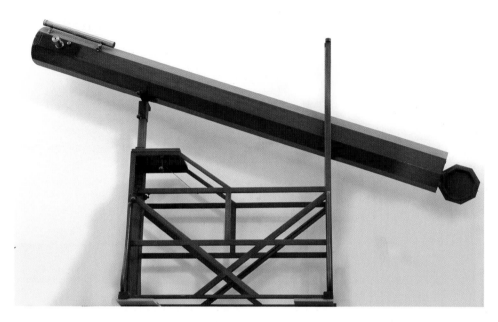

Plate 3. One of the five 10-foot reflectors ordered by King George from William in October 1783. The king presented it in 1786 to the Duke of Marlborough to thank him for his hospitality during a royal visit to Blenheim Palace. Courtesy of the Whipple Museum for the History of Science, University of Cambridge.

Plate 4. William painted in oil on canvas in 1785 by Lemuel Francis Abbott (1760 or '61–1802). By the early 1780s Abbott had established a busy portrait practice in London, concentrating on men from the professional classes and the navy (including Admiral Nelson). In December 1784 he completed a portrait of William's friend William Watson, who was so delighted that he urged William to follow his example. The Herschels were in turn equally delighted, so much so that after William's death in 1822, his son John commissioned for his widowed mother a miniature based on Abbott's portrait. In 1798 Abbott was certified insane. © National Portrait Gallery, London.

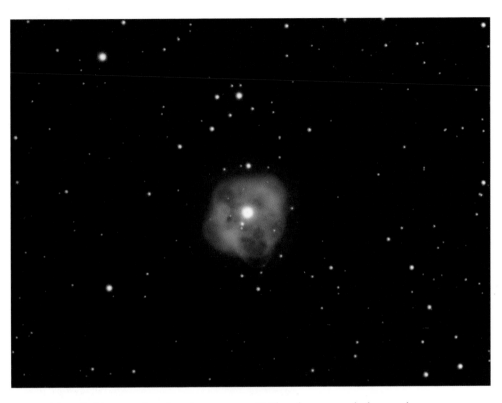

Plate 5. "A most singular phaenomenon!" The planetary nebula now known as NGC 1514, which William encountered on November 13, 1790. The nebula is near enough to Earth for William to see the central star, which he thought was condensing out of the surrounding nebulosity. This convinced him that "true nebulosity" did indeed exist, and so the principal question that had motivated the construction of the 40-foot reflector had been answered. Image courtesy of Dietmar Hagar, FRAS, © www.stargazer-observatory.com.

Plate 6. William's wife Mary in March 1805, miniature on ivory by John Keenan. Keenan was born in Ireland, and from 1802 he worked in the Windsor area, becoming court painter to Queen Charlotte in 1809. Between 1791 and 1815 he exhibited some sixty pictures at the Royal Academy, nearly all of them portraits. Herschel Family Archives.

Catch

What a strong Sulpherous scent proceeds from this meteor.

The Female Philosopher smelling out the Comet

Pub Feb. 1, 1790 by R. Hawkins N.53 old Lamson? St. John?

Plate 7. Etching dated February 1790, and possibly by R. Hawkins, showing "The Female Philosopher"—undoubtedly Caroline—"smelling out the Comet"—but apparently she is mistaking the comet for a meteor! Caroline had found a comet the previous month. Draper Hill Collection, the Ohio State University Billy Ireland Cartoon Library & Museum.

Plate 8. Watercolor of the 25-foot reflector made by William in 1796–98 for the King of Spain, from the instructions for assembly that accompanied the instrument when it was shipped to Madrid in 1802. This was the most successful of William's large reflectors, with greater "light gathering power" than the 20-foot but easily maneuverable despite its size. In 1808 the mounting was destroyed by Napoleonic troops, but the optics and the instructions for assembly survive. A full-scale replica of the telescope has been constructed recently in northern Spain. Courtesy of the Observatorio Astronómico Nacional, Madrid.

Plate 9. One of the numerous 7-foot reflectors that William made for commercial sale. Photograph by J. Karpinski, Dresden, courtesy of Staatliche Kunstsamm-lungen Dresden, Mathematisch-Physikalischer Salon.

The Georgian Planet with its Satellites.

Plate 10. Pastel portrait, 1794, of William by John Russell (1746–1806), Crayon Painter to King George III. Russell was a leading portrait painter of the day, and he exhibited at the Royal Academy from 1769. He was elected to the Academy in 1788. Russell had a keen interest in science, and devised lunar maps and globes. Herschel Family Archives.

Plate 11. Portrait in oils of John Herschel by Robert Muller in 1799, when John was seven, with Windsor Castle in the background. Muller was born in 1773. He was remarkably young when he enrolled as a student at the Royal Academy in 1788, and he first exhibited there the following year. In all he exhibited twenty-eight portraits at the academy between 1789 and 1800, among them "Master Herschel" in 1799. Muller had previously painted Mary Herschel's mother, Mrs. Elizabeth Baldwin, who died in 1798. Herschel Family Archives.

Plate 12. Portrait in oils of John Herschel in 1833, by Henry William Pickersgill (1782–1875). By permission of the Master and Fellows of St John's College, Cambridge.

Plate 13. Portrait in oils of Caroline in 1829, a few days before her seventy-ninth birthday, by Melchior Gommar [otherwise Maerten Franz or Martin François] Tieleman [or Tielemans or Tielemann] (1784–1864). Tieleman was born in Lier, Belgium, and studied in Antwerp and then with Jacques-Louis David in Paris. In 1815 Hanover was raised in status to a kingdom, and when the Duke of Cambridge, son of George III, was made viceroy the following year, he appointed Tieleman court painter. Caroline sat eight times for her portrait, which Tieleman delivered to her on March 26, the day before he left Hanover to take up a professorship near Antwerp. Herschel Family Archives.

Plate 14. Watercolor by Henrietta M. Crompton (1793–1881) from the 1840s showing the third Earl of Rosse supervising the positioning of the 6-foot mirror in the tube of the "Leviathan of Parsonstown." To the left is the 3-foot reflector and, far left, Birr Castle. © The Birr Scientific and Heritage Foundation, courtesy of the Earl of Rosse.

Plate 15. Caroline Herschel's tomb, in the Gartenkirchhof on the Marienstrasse, Hanover. In the earth beneath lies the body of her mother Anna, and below Anna is the body of Caroline's father Isaac. The inscription, which Caroline herself composed, makes mention of Isaac but none of Anna, who therefore lies today in an unmarked grave. Photograph courtesy of Owen Gingerich.

Messier being unable to search at the time because he was still recovering from a fall into an ice cellar.

Caroline's Index to Flamsteed's British Catalogue

Maskelyne's admiration for Caroline increased still further as the century drew to a close. When observing, she and William had always relied on John Flamsteed's great British Catalogue of three thousand stars. Published in 1725, this was thought to be the infallible guide to the night sky. But sometimes William came across puzzling discrepancies—stars that had apparently appeared or disappeared or had altered in brightness. At first he assumed these changes were genuine and attempted to contrive explanations for them, but it slowly dawned on him that the British Catalogue was not infallible after all.

The catalogue itself formed volume 3 of Flamsteed's *Historia coelestis*, and the individual observations on which the catalogue entries were based were in volume 2. The question was, had the observations in volume 2 been, without exception, faithfully and accurately assembled into the catalogue? Or were there scribal and printer's errors? Unfortunately Flamsteed had provided no index that would have allowed the user to go back and check a catalogue entry in volume 3 against the relevant observations in volume 2. And so, in the autumn of 1795 William "invited" Caroline to supply such an index. The task was straightforward enough—but it required twenty months of dedication, perseverance, and above all meticulous accuracy. Fortunately these were qualities that Caroline had in abundance. While compiling the index she found hundreds of errors, as well as over five hundred stars that Flamsteed had overlooked—the British Catalogue should have included three and a half thousand stars.

Her work added immensely to the reliability and usefulness of the catalogue, and in a striking commendation of her labors the Royal Society, at Maskelyne's urging, published Caroline's catalogue at its own expense. The volume contains seventy-six folio pages of symbols and must have been a nightmare to typeset and the same to proofread. William contributed a patronizing foreword, which even his greatest admirers may find unpalatable:

> And I may add, that by inspecting the work as it proceeded, and
> looking over all cases which seemed to require more of the habits of

an astronomer than she has been in the way of acquiring, I have endeavoured, as much as I could, to prevent errors from finding their way into the work.

If there is a single slip in the thousands of pages of meticulous records that Caroline copied out during her half-century in England, it has yet to be identified.

Caroline was delighted to have her work published by so august a body as the Royal Society and wrote to Maskelyne to thank him:

> But your having thought it worthy of the press has flattered my vanity not a little. You see, Sir, I do own myself to be vain, because I would not wish to be singular; and was there ever a woman without vanity? —or a man either? only with this difference, that among gentlemen the commodity is generally stiled ambition.

Maskelyne knew Caroline well enough to comment that he personally would have found the list of omitted stars more helpful if they had been arranged to suit the needs of the observer rather than by constellation. Without hesitation Caroline sat down and rewrote the entire list—running to twenty-five pages of numbers—in the format he wanted.

No wonder that Maskelyne thought the world of her and several times had her to stay at Greenwich. She would come with some job in mind, but Maskelyne secretly disapproved of the Slough regime of uninterrupted labor and would devise a program of entertainments that she could not avoid. But it was against Caroline's nature to fail in what she saw as her duty, and after she was at last allowed to retire to her room, she would sit up half the night working at her desk. On one occasion she tells of a visit

> I had intended to spend at Greenwich for the purpose of copying the memorandums from my brother's second volume of Flamsteed's Observations into Dr. Maskelyne's volume. But the succession of amusements, &c., &c., left me with no alternative between contenting myself with one or two hours' sleep per night during the six days I was at Greenwich, or to go home without having fulfilled my purpose.[4]

On one occasion, however, Maskelyne took matters into his own hands. Having lured Caroline to Greenwich on some pretext, he then took her on a visit to the observatory of William's old ally Alexander Aubert, after which he carried her off for a week to Sir George Shuckburgh, another leading amateur astronomer. He was sure, he told William, that for once her brother could manage without her for a few days.

10

1788–1810
"The Most Celebrated of All the Astronomers of the Universe"

While William was courting Mary, a letter arrived at Slough from Jérôme de Lalande in Paris. It was addressed "A Monsieur Herschel, le plus célèbre de tous les astronomes de l'univers, Windsor."[1] Monsieur Herschel had been a professional astronomer for just four years and nine months.

"A knowledge of the construction of the heavens," William later wrote, "has always been the ultimate object of my observations." By the end of the 1780s he had published three great papers in which he explored the cosmos, portraying it not as the unchanging mechanism of God the Clockmaker, but as an arena in which bodies great and small pass from youth to old age as gravity works its magic. He would return to the subject in the 1810s in his last four major contributions to *Philosophical Transactions*, where he would illustrate his vision of an evolving universe with innumerable examples taken from his great catalogues. But his discovery in 1790 of decisive evidence for the existence of "true nebulosity" brought closure to the controversy that had exercised his mind since his first days as an amateur observer and that had motivated his construction of the 40-foot reflector. In the 1790s and 1800s, therefore, he felt free to devote more of his effort to other topics, not least the bodies of the solar system. It was the solar system that still preoccupied most other astronomers, and with his fine telescopes, William was well placed to contribute to this traditional domain of astronomy. Indeed, half of his total publications concern the Sun and the planets and comets.

The Search for Uranian Moons

The Georgian Star had of course a unique place in his affections, and once it was proved to be a major planet, he had been keen to discover whether it had moons, and if so, how many. The difficulty was that the planet takes no less than eighty-four years to orbit the Sun, and in the 1780s it was slowly moving in a region rich in telescopic stars that were difficult to distinguish from possible moons. But hope revived early in 1787, when he reexamined a number of nebulae but this time using a new configuration (now known as "Herschelian," or "front view") that he had adopted the previous autumn for his 20-foot reflector; to his delight he found the nebulae appeared much brighter than in the past.

In the standard Newtonian construction, starlight is reflected from the curved main mirror at the bottom of the tube, back up to the little flat mirror near the top. This then reflects the image sideways to the eyepiece. But William found that in this second reflection he was losing nearly half the precious light, and so as early as 1776 he tried to see if it was possible to dispense with the second mirror: instead he peeped directly down the tube through an eyepiece positioned at the edge of the opening at the top. Unfortunately, in a small reflector with a tube of modest size, the observer's head then blocks much of the light, and so there might be more loss than gain in adopting this "front view." But in a big reflector of wide aperture, such as the 20-foot (figure 21), the observer's head would cast a relatively small shadow down the tube. True, the main mirror would have to be tilted very slightly, so that the light converged at the edge of the opening at the top of the tube instead of at the center, and some minor distortion would result; but this was a price worth paying.

Encouraged by the increase in the brightness of the nebulae, William reobserved the region of sky immediately around his planet, including "some very faint stars whose places I noted down with great care." Were there, hidden among them, one or more moons of Uranus? If so, these moons would eventually betray themselves by accompanying the planet as it moved slowly among the genuine stars. Within weeks William was certain that Uranus had two moons.[2] And by the following year, 1788, he had determined the time these moons take to orbit their parent planet, to within a few minutes of the modern values. As he had also determined that the planet as seen from Earth had an apparent diameter of just under four seconds of arc, he was now able to use Newtonian mechanics to calculate Uranus's mass, volume, and density.

But to William a mere two moons did not seem enough for this major planet of his—Jupiter was known to have four, and Saturn seven—and

Figure 21. William's 20-foot reflector in its prime, from an engraving he published in 1794.

so, in the closing years of the century, he put a lot of time into the search for more. By 1797 he had persuaded himself that he had found another four, which would have brought the total to six. Uranus does indeed have numerous moons, as the *Voyager 2* spacecraft discovered, but it seems that none of them matches any of William's four: for once he had deceived himself.

The Structure of the Sun

Nevil Maskelyne had long ago taught William not to mix science and religion, and in his great papers of the 1780s, William had been careful never to so much as hint at his conviction that every star and planet in the universe is peopled with intelligent beings. In private he—or perhaps Alexander—was less discreet, for the *Bath Chronicle* in April 1793 informed its readers that William "is now said, by the aid of his powerful glasses, to have reduced to a certainty, the opinion that the moon is inhabited." Indeed, "he has distinguished a large edifice"; this building, it seemed, was comparable in size to St Paul's Cathedral. Not only that, but he "is confident of shortly being able to give an account of the inhabitants." One wonders if these remarks were penned on April Fool's Day.

It was in his theory of the physical nature of the Sun (and therefore of stars in general) that William's beliefs most evidently impacted on his science. Surely it was impossible for any form of life to withstand the solar heat at close quarters? William's response, set out at length in *Philosophical Transactions* in 1795, was to argue that the core of the Sun is cool and solid, like that of a large planet. Indeed, the Sun actually is "a very eminent, large, and lucid planet." Its solid surface is surrounded by several "elastic fluids" (gases), one of which is luminous; sunspots are holes in this luminous fluid that allow us to see the dark surface of the Sun. The inhabitants of the Sun are able to survive because rays from this luminous fluid produce heat only when they enter an appropriate medium, such as the Earth's atmosphere. After all, the highest mountains on Earth are covered with ice even though they are exposed to sunlight for long periods and have no clouds to shield them. In this paper, for once only, he talked freely in print about lunar, solar, and stellar inhabitants, and for once his prestige allowed him to get away with it.

In a second paper, read to the Royal Society in April 1801 (a month after his supposed death had been announced in the *Morning Herald*[3]), he modified this scheme a little and argued that the planetary Sun is surrounded by inner, dark clouds that shield the inhabitants from the outer, fiery ones. He also wondered whether the heat we receive from the Sun varies over time, perhaps linked with fluctuations in sunspots. The Sun's heat of course affects harvests on Earth, and so he studied the prices of wheat at Windsor to see if there was any correlation of these prices with records of sunspots. He listed the prices for five periods when there had been few or no spots on the Sun, and in a paragraph he interpolated in his draft at the final stage, he drew the tentative conclusion that in these periods "some temporary scarcity or defect of vegetation has generally taken place." Watson was aghast at such speculations. As he forecast, William was ridiculed for his pains. The Scot Lord Brougham, writing in the newly founded *Edinburgh Review* (a journal that loathed the English and especially those of German origin), described the suggestion as a "grand absurdity. . . . Since the publication of Gulliver's voyage to Laputa, nothing so ridiculous has ever been offered to the world."[4] William might have got a more sympathetic hearing from a modern reader.

Infrared Rays

In order to observe the Sun without damaging his eyes, William had of course to resort to a variety of colored filters, some of which were made for him by Alexander. While William was experimenting to find the most suitable, he formed the impression that the heat from the Sun reaching his eyes varied with the color of the filter, and he set up an apparatus with a prism and a number of thermometers to test whether this was indeed so. He found that his hunch was correct: red light from the Sun produced more heat than green, and much more than violet. The heating increased toward the red end of the visible spectrum.

It might even be that the heating effect did not end there. And so it proved: beyond the red there were rays from the Sun that conveyed heat but no light. William announced his discovery of "infra-red rays," as we know them, in two papers published in *Philosophical Transactions* in 1800. Sir Joseph Banks thought this the most important of all his many contributions to science.

"Asteroids" or "Planetoids"?

Uranus had offered unexpected vindication of what we know as Bode's Law. Take the first six integers in which each is twice its predecessor: 1, 2, 4, 8, 16, 32. Multiply these numbers by 3 (giving 3, 6, 12, 24, 48, 96), add 4 to each (so we get 7, 10, 16, 28, 52, 100), and now begin the sequence with a 4: 4, 7, 10, 16, 28, 52, 100. It looks like numerology and probably is. But there had long been keen interest in the layout of the solar system—in the distances of the various planets from the Sun. A couple of centuries before, Kepler had asked himself what had motivated God the Geometer to choose these distances rather than others, and in a nesting of six spheres and the five regular solids, he had found an answer that satisfied him (but no one else). Then, in 1766, the German astronomer Johann Daniel Titius pointed out that the distances were (almost) proportional to the first four and the last two of these seven numbers 4, 7, 10, 16, [28], 52, 100. His idea was taken up and popularized in 1772 by the nineteen-year-old Johann Elert Bode, hence the name of the "law."[5]

Bode's Law focused attention on the huge gap between Mars and Saturn, where there was no known planet corresponding to the number 28.

Was the gap occupied by a planet as yet undiscovered? It was while astronomers were ruminating on this question that William came across Uranus, and Uranus's orbit proved to correspond well to the next term in the sequence, $(64 \times 3) + 4 = 196$. It was amazing then and it is amazing now. It lent a whole new plausibility to the "law," and near the turn of the century Baron Franz Xaver von Zach, court astronomer at Gotha, organized a meeting of like-minded observers, all eager to see if there was indeed a planet in the gap. They decided to divide the zodiac—the zone where the Sun and planets are to be found—into twenty-four sections, each section to be assigned to an astronomer who would act as a "celestial policeman" and keep a lookout for any suspicious characters seen loitering in his district.

Meanwhile at Palermo, the southernmost of the European observatories, Giuseppe Piazzi was at work compiling a star catalogue, blissfully unaware that he was one of those to be recruited into the celestial cops. His careful method of working required him to measure the positions of his stars on two different evenings, and on New Year's Day 1801, he measured a faint "star," which on reexamination proved to have moved. It was therefore not a star but a member of the solar system.

Piazzi tracked Ceres, as he called it, for several weeks, until he lost it in the glare of the Sun. He had observed the object for only a tiny fraction of its orbit, and it was in danger of being permanently lost. On September 2 Piazzi wrote to William, appealing to him to search for it in his great telescopes; and at the end of October William replied to say that he was on the job (and hinting that Piazzi was a fool not to have notified him much earlier, when the Italian first discovered the object!). William naturally supposed that he would be able to identify Ceres by its visible disk, as he had the Georgian Star two decades earlier; but in this he had no success.

Fortunately the talented young German mathematician Carl Friedrich Gauss succeeded in overcoming the problem of the fragmentary nature of Piazzi's observations. He was able to predict where the object would reappear, and on the last night of the year, it was located by von Zach. Ceres's distance from the Sun corresponded to the missing term 28 in the sequence of Bode's Law, and so far there was no reason to doubt that it was the major planet that had long been expected in this gap between Mars and Jupiter. But William still found it impossible to identify any object in the region that had a planetary disk, and it was only when Maskelyne gave him an exact location early in February 1802 that he was able to find it. But its disk proved to be so minute that he had great difficulty in distinguishing it from a star. His preliminary conclusion was that Ceres's diameter was not much more than half the diameter of the Moon.[6] This was a very strange planet indeed.

Soon after, a similar object was found by an amateur astronomer of Bremen, Heinrich Olbers, who named it Pallas. William tracked it down and decided that it too was tiny—so tiny that the sphere of little Mercury had space for 31,000 such objects.

It was time for him to go public with his measurements, which would astonish astronomers home and abroad. But what term should he use to designate these little objects, which were neither planetlike nor cometlike and so in his view formed a new species of celestial body? On April 18, 1802, he wrote to Banks to ask if the Royal Society had settled on a term (they had not), and a week later he consulted William Watson, who had been responsible for the name of the Georgium Sidus. He told Watson:

> they are extremely small, beyond all comparison less than planets, & move in oblique orbits. . . . I surmise (again) that possibly numbers of such small bodies that have not enough matter in them to hurt one another by attraction, or to disturb the planets, may possibly be running through the great vacancies, left perhaps for them, between the other planets especially Mars and Jupiter.

He concluded: "However, you will give it a thought, and if two or three names could be proposed it would give me some choice."

This was the kind of challenge Watson relished, and he had numerous suggestions to offer, all diminutives of "planet": *planetel* (as a cockerel is a small cock); *planetet* (as a baronet is a minor baron); *planetkin* or *erratikin* (as a lambkin is a small lamb); or even *planetule* (as a spherule is a little sphere). *Planetling*, by analogy with "duckling," he disliked.

In May William's elaborate analysis of the physical characteristics of the two objects was read to the Royal Society. He argued that they formed a class intermediate between the planets and the comets; and, ignoring Watson's suggestions, he proposed that because they looked so much like stars (in Greek, *aster*), they might be termed *asteroids*. One of his many objections to calling them planets was that this would place two such bodies in the Mars–Jupiter gap, and so violate Bode's Law, which provided for only one. But Lord Brougham in the *Edinburgh Review* declared that William's "passion for coining words" was "a weakness wholly unworthy of him. The invention of a name is but a poor achievement in him who has discovered whole worlds."[7] He went on to imply that his motive was to remain the only discoverer of a planet. Gauss was equally critical: in a letter to Olbers he declared William's outlook to be "unphilosophical."[8]

The following month Olbers wrote to William suggesting that Ceres and Pallas might be fragments of a major planet that had disintegrated long ago, perhaps by explosion or the impact of a comet. This of course would salvage Bode's Law. A third such body (Juno) was found in 1804 by the German astronomer K. L. Harding, and Olbers found yet another (Vesta) in 1807. William deemed these too to be asteroids. But Piazzi preferred the name *planetoid*, for the bodies were in the planetary system and "I put it to you, the name of Asteroids seems to me more appropriate for little stars."

It is William's name that has stuck, but his motives continued to be questioned. In 1812 the Scot Thomas Thomson, in his *History of the Royal Society*, said that the four planetary bodies in the Mars–Jupiter gap were

> so small that Dr. Herschel refuses to honour them with the name of planets, and chuses to call them asteroids, though for what reason it is not easy to determine, unless it be to deprive the discoverers of these bodies of any pretence for rating themselves as high in the list of astronomical discoverers as himself.

Investigating Saturn

William published papers on most of the planets of the solar system, but his enduring fascination was for Saturn. It was the first object he had viewed when he opened his first observing book, and his discovery in 1789 of the sixth and seventh moons had provided a triumphant entry for the 40-foot on the astronomical scene. In the years that followed he dedicated no fewer than six complete papers to the planet, containing meticulous studies of almost every aspect of Saturn, its moons, and its ring. So elaborate are William's investigations that the standard book on the history of

Saturnian studies[9] devotes two complete chapters to them, and by the time
he had finished he had a knowledge of the planet and its appendages that
was unparalleled in its detail. Saturn was, he concluded, a

> magnificent globe, encompassed by a stupendous double ring : at-
> tended by seven satellites : ornamented with equatorial belts : com-
> pressed at the poles : turning upon its axis : mutually eclipsing its
> ring and satellites, and eclipsed by them : the most distant of the
> rings also turning upon its axis, and the same taking place with the
> farthest of the satellites : all the parts of the system of Saturn occa-
> sionally reflecting light to each other : the rings and moons illumi-
> nating the nights of the Saturnian : the globe and satellites enlight-
> ening the dark part of the rings : and the planet and rings throwing
> back the sun's beams upon the moons, when they are deprived of
> them at the time of their conjunctions.[10]

The Baleful Fascination of Newton's Rings

William's researches now took an unfortunate turn, for he became obsessed
with the elucidation of "Newton's rings," the concentric colored rings that
appear when two pieces of glass are pressed together. As far back as 1792
he had borrowed a couple of lenses by Huygens from the Royal Society
and had begun experiments with them. Then his long-standing correspondent Patrick Wilson retired from his professorship at Glasgow University
and moved to London. It could be said that scientifically William fell into
bad company. He met regularly with Wilson to discuss Newton's rings
and allied problems, meetings that they termed "philosophical chapters."
For a time William came close to abandoning astronomy altogether, as he
made endless experiments on rings, assembling for *Philosophical Transactions* three papers of enormous length, which he submitted in 1807, 1809,
and 1810. The Royal Society's committee that vetted papers for publication was most unhappy about them, and William's friends were embar-
rassed, but such was his scientific eminence by this stage in his career that
the papers were reluctantly allowed into print. Later, when John planned
an edition of his father's publications, he intended to omit these three,[11]
and the edition that eventually appeared in 1912 includes the ring papers
only for completeness.

Reflectors Great and Small

The profits from William's trade in telescopes had helped pay for the experiments he needed to make in preparation for the construction of the 40-foot, but once he had married into money, he had ample funds at his disposal and no need to continue making instruments for sale. Yet his appetite for trade was undiminished. In part this was because, like the modern supertanker, once on course he was not easily diverted. He may have hoped for confirmation of his observational claims from the few fellow astronomers who bought one of his larger instruments, for not everyone was prepared to take on trust William's descriptions of what he alone could see. But one suspects the primary motivation was his delight at the prestige this work brought him. The crowned heads of Europe begged for the privilege of being allowed to buy a Herschel reflector, for if William turned them down, they had nowhere else to go. Such letters from kings and emperors massaged his ego. He was telling no more than the truth when writing in March 1794 to the representative of the Empress of Russia: "I have only to add that I feel myself extremely flattered by the notice of her Imperial Majesty." The empress was invited to choose from a list of nine reflectors, ranging from a small 7-foot with mirror 6¼ inches costing 100 guineas (such as the one shown in plate 9), to 25-foot, 30-foot, and 40-foot reflectors at 4,000; 6,000; and 8,000 guineas, respectively (these larger instruments at half-price if the wooden stand was to be provided by the purchaser).[12] William had earlier offered potential purchasers three different grades of instrument—if the 7-foot came with "heigh powers and an apparatus for day objects" the price would be 150 guineas, while "The same with micrometers and every other advantage for viewing the Sun and planets" would cost 200 guineas[13]—but he found little interest in the more expensive versions, and they soon disappeared from his price list.

William never found a buyer for the 30-foot or the 40-foot, fortunately we might think, but around the turn of the century the King of Spain bought a 25-foot (plate 8) as well as two smaller instruments. The mirror of the 25-foot had a diameter of 2 feet rather than the 2½ feet of the price list; and perhaps the modesty of this scaling-up of William's own 20-foot was well advised, for the 25-foot was to prove the most successful of all his great telescopes.

As usual with his major instruments, William cast two mirrors, for sooner or later the one in use would tarnish and need repolishing, and then it could be replaced for a while by its twin. The castings took place in August 1796, and he ground and polished throughout the winter and early spring of 1797, and again in the first months of 1798. As he did so he made regular observations to test the mirrors, an experience that gave him much satisfaction. Eventually, in April 1801, the instrument was disassembled, but it would be the following year before it was packed into fifty-two boxes and loaded onto the Danish brig *Juana* for shipment to Bilbao, from where it would be transported overland to Madrid. Sadly, the mounting was destroyed in 1808 by Napoleonic troops, but the two mirrors and the drawings and elaborate instructions for assembly survive in Madrid to this day. A full-scale replica has recently been constructed, and so it is possible once more to admire William's ingenuity as a builder of great telescopes.[14]

William never in his life traveled much further south than Paris, and as an observer he was never able to view the skies that are forever below the horizon at Slough. But he envied the clear nights enjoyed by astronomers in the Mediterranean regions, and he twice made attempts to go there himself. The first we know only from a cryptic remark in a letter from William Watson dated August 25, 1784: "Has the scheme succeeded of going into Italy?" William made his second attempt seven years later, when Sir William Hamilton, the ambassador to the kingdom of Naples, visited London, and William called on him there and declared that he was keen "to make some observations on the Planets in the Climate of Naples." Would the King of Naples care to buy one of William's 20-foot reflectors?—in which case "Dr. H will get one made and come to Naples to put it up and observe some time with it."[15] But nothing came of the proposal.

Britain's Early-Warning System

All of William's clients intended to use their telescope to study the heavens, except one. He was Robert Smith, first Lord Carrington and a crony of the prime minister at the turn of the century, William Pitt the younger. The telescope he commissioned—a 7-foot at the usual price of 100 guineas—was to provide an early-warning system in the event of a French invasion.

It was to be installed at Walmer Castle, overlooking the Straits of Dover, and it was to keep an eye on Britain's traditional enemies from across the water.

The instrument arrived at the castle in the autumn of 1799 in kit form, with a dozen pages of instructions for assembly written in Caroline's hand. The politicians were baffled. As William's friend, the poet Charles Burney, later told him, "no one had sufficient skill to put it together, and render it useful in examining the French coast and intermediate sea, for which it was originally intended." Burney was unlucky enough to be at the castle, and on entering a room he found there a group that included George Canning, the undersecretary for foreign affairs. The men were gathered round the nation's early-warning system, which was spread across the room in pieces, and they were trying to puzzle out from Caroline's elaborate instructions how to put it together.

Burney's arrival proved very welcome, for as the author of an astronomical poem, he must surely know how to assemble a telescope. The poet was dismayed, for he had not the slightest idea what to do; but rescue was at hand. Before Burney had had time to read more than a few lines of the instructions, visitors arrived, and in the ensuing confusion he was able to slip the pages back in a drawer, "and to say the truth, without much reluctance, for I doubted my competence."

Before he could be summoned for a second attempt, Burney sent an anguished plea to William for help. Was there someone in the Walmer area who knew about assembling telescopes? If not, would he please send fuller instructions—which William did.[16]

A couple of weeks later the prime minister himself was at Walmer, trying his hand at assembling the telescope. Pitt had mastered the first two of the six steps prescribed by William, but was baffled as to what to do next.

> Mr Pitt perused your instructions with satisfaction; & when he came to the 6 short rules laid down—to the first 2 he said:—"we are right thus far—but the putting on the small speculum [mirror] has puzzled us"—the rest he seemed to think not difficult. And said that he had left a person at work, who seemed to understand the business.[17]

What became of the telescope no one knows.

11

1792–1822
The Torch Is Handed On

A Son and Heir

On March 9, 1792, Mary gave birth to a son, John. At last the spinster Caroline had a nephew on whom she could dote and toward whom she would one day admit to feeling "motherly."[1] Paul Pitt, Mary's surviving son,[2] wrote to William expressing pleasure "on the birth of a son to you and a brother to me,"[3] but within a year Paul died, probably of tuberculosis (or consumption, as it was then termed). John would always regret being deprived in this way of the companionship of an elder brother. Caroline believed that John had narrowly escaped a similar fate, for his wet nurse also died of consumption.[4]

William, who was now well into his fifties (plate 10), took his paternal duties lightly. When John was less than three months old, William departed on a leisurely tour with his friend the Polish count, Lieutenant-General Jan Komarzewski, one of John's godfathers. Both men were fascinated by machinery, and together they toured Britain's industrial heartland. William could converse with leading engineers on terms of equality; indeed, in 1793 he would be a witness for James Watt in the case of *Watt v. Bull*. During the tour he made over one hundred elaborate drawings of the devices they were shown, with extensive notes, and these form a valuable record of the industry of the time.

The ostensible purpose of their journey was to visit Glasgow. There both men were granted the freedom of the city, and the provost proposed the health of the new burgesses. The same evening the Principal of Glasgow University presented William with his diploma as honorary Doctor of Laws. Then the two friends turned for home. On the way they called to see the ailing John Michell, who years ago had questioned William's use of double stars as a way of measuring stellar distances and who was now rector of the parish of Thornhill in Yorkshire. They reached Slough after an absence of seven weeks, and William was then home a mere couple of days, barely long enough to get to know his infant son, before he was off again, this time to Cornwall. William achieved so much in life partly because he always put his own interests first.

Even before John was old enough to travel, William and Mary began to indulge in regular (and often extended) holidays although William would sometimes salve his conscience by taking his "travelling 7-foot" reflector with him. When John was only fifteen months old, his parents went off by themselves on an eight-week tour that took them to Thornhill. Michell had died that April, and William paid his son-in-law thirty pounds for the remnants of his telescope, which with its 30-inch mirror (now broken into four pieces) had rivaled William's in size though not in quality. "I gave the person to pack it up, 1½ Guinea & ½ Guinea to the Carpenter for his trouble." He was fascinated to have at last the chance to examine the work of his rival, and there is a certain smug satisfaction in his criticisms: "Its construction is what I cannot approve."[5]

John was taken on his first tour in August 1797, when he was five. Soon afterward, Mary's niece Sophia Baldwin, "a sweet timid amiable girl," joined the family as companion to Mary, and she went along when William, Mary, and John set off on their 1798 tour. They first visited their old haunts in Bath, where they took John to a French puppet show. William's curiosity had now moved on from machinery to geology, and he was particularly intrigued by the marine fossils sometimes to be found at heights far above the present sea level. He and his brother Alexander left the others for a day so that they could examine the "mountains" near Bath. "They contain shells, but not in abundance." The party then made their leisurely way to William's favorite seaside resort of Dawlish on the south Devon coast, where his old ally William Watson had a holiday home.

On the return journey, William explored a hill between Exeter and Honiton. Although it was now far inland, and rising to one thousand feet above sea level, "the work of the sea shore is clearly to be perceived." William envisaged the sea level slowly rising, with sediment continually washing down onto the levels below. Then, "perhaps thousands of centuries after," the process was reversed. Not everyone thought the world was created in 4004 BC.

In the tours that followed, the party nearly always consisted of William, "Mrs Herschel, Miss Baldwin, John Herschel," but never Caroline—she was invariably left at home to "mind the shop." After all, she was being paid by the Crown to assist William, if only by showing the telescopes to visiting aristocrats in his absence. The party always traveled in style. In 1809 their eight-week tour of the north of England (in their own coach and horses) cost them £188 17s, almost the whole of William's pension for the year, and the following year and again in 1811, William set off with no less than £400 in his cash box.

Meanwhile Caroline and her little nephew formed a close bond, and Mary had no hesitation in leaving her in charge of him for weeks on end. John showed an independent mind from the start. When he was still a toddler, the Herschel home was "extended." It was the custom on such occasions for the builders to place a coin for luck at the base of the cornerstone, and when the time came, Caroline lifted John up to perform the ceremony. The infant, showing good sense beyond his years, objected to this waste of money, and finally complied with his aunt's wishes only under protest.[6]

It was from Caroline that John received his introduction to astronomy. She used to amuse the child by showing him the constellations in Flamsteed's Atlas. The fearsome representation of the constellation Cetus—the Whale—especially fascinated him, and when his nurse brought him to her he would demand, "Aunty shew me the Wail!"[7]

The household in which John grew up was unorthodox in the extreme. He was an only child in a world of adults, most of them middle-aged. His father and aunt often worked at night; but in the daytime, as his daughter later put it,

> [t]here were plenty of exciting operations for a small boy going on at home. In one workshop he might find his extremely skilful Uncle Alexander doing a bit of turning on the cumbersome foot-lathe or busy at another bench cutting a brassfitting for an instrument. At the furthest end of the walled garden was a foundry, a glorious sight, with flames and flying sparks, where sweating men hammered at the glowing metal whilst boys no older than himself pulled the rope for the bellows.

And if John was taken to see the Upton Farm that his mother owned, they would pass "several narrow, low, thatched barns where his father's carpenters worked at the wooden parts of the long telescope tubes and frameworks."[8]

John (plate 11) was to have an unorthodox schooling. His mother had more money than sense and "became over solicitous to keep him from the children of toadies and sycophants hanging around the Regent and the Court at Windsor."[9] His education could be described as force-fed. One of John's contemporaries later remarked, "He never was a child, he hadn't played with other boys or joined in other children's games."[10] When he was five he was sent away to a school run by a Mr. Atkins, and the unfortunate child "came home Sundays and others to receive instruction by Teachers in writing, Arithmetic, Music, Geography, etc."[11] He was sent next for a few terms to Mr. Ball at Newbury. Newbury is over thirty miles from Slough, but the coaches that passed through Slough on their way from London to Bath all stopped there. No doubt he would be met by his godmother, Susan White, a local resident, and she would keep an eye on him while he was at the school.

Then, on January 30, 1800,[12] when he was nearly eight years old, John entered Eton College, today the most famous of all England's public (that is, private) schools but which an old Etonian friend of William's later described as "that seat of vapid lounging."[13] After the Easter break, on May 1, Caroline took him back to "his boarding dame, Mrs Howard." But he was at the school for only a few days more, for when Mary was riding through Eton village, she was appalled to see her delicate son "stripped and Boxing with a great Boy."[14]

John was removed forthwith and before the end of the month was installed as a border in a private school in Hitcham House, a Georgian property five miles west of Slough, where education was primarily in classical languages and which catered to "the Nobility and Gentry."[15] It was run by Rev. George Gretton, a former Fellow of Trinity College, Cambridge, and a Doctor of Divinity, who in 1803 was to become rector of nearby Hedsor. At Cambridge Gretton had twice been awarded one of the prizes for Latin composition sponsored by the members of Parliament for the university. Education at Hitcham House was very different from the basic reading, writing, and arithmetic offered to children of farm laborers in the normal village school, and John's parents were happy to pay Gretton fees that represented more than half of William's salary.

But early in 1804 disaster struck: Hitcham House was burned down and had to be demolished.[16] While Gretton searched for alternative premises, John was sent to a private tutor, a Mr. Luscombe, who lived in the village of Clewer near Windsor, a couple of miles from the Herschel home. In

January 1805 John was back with Gretton, and he remained at the school until 1809, when he entered St. John's College, Cambridge. Caroline remembered with pride one particular occasion at Gretton's, the speech day on July 14, 1806, when she attended along with William, Alexander, and Dietrich, who was on a visit from Hanover. John, aged only 15, made the third speech and "surprised us all."[17]

Even so, Gretton alone was not good enough for Mary, and so she supplemented the curriculum with private home tutors, notably a Scot named Rogers, who traveled all the way from Edinburgh to Slough in October 1806 at the cost of four pounds. John would go to Gretton several days each week, and study with Rogers on the remaining days. Rogers introduced John to mathematics and French, and this arrangement lasted for a year. Then, in 1808, it was decided John should learn German; a Mr. Hausmann was recruited to teach him for a fee of three guineas per annum, and he continued to do so until John went to Cambridge.

But meanwhile, in Slough all was not well.

Caroline Quits the Herschel Home

How Caroline's career as a singer would have developed if she had long ago accepted the invitation to sing at Birmingham we can only speculate. Certainly the history of astronomy profited hugely when she committed her future to William. But it is also certain that astronomy suffered from her abrupt decision in October 1797 to quit her comfortable and convenient cottage in the grounds of William's home and move into rented accommodation.[18]

Just what provoked this extraordinary step we shall never know, for Caroline later shamefacedly destroyed her diary for the period. Mary Herschel was the gentlest of souls, and William would never knowingly have offended his devoted little sister. But one suspects that either William or Mary in an unguarded moment used words that reminded Caroline of her status as their lodger. Until recently she had never had money of her own and so had always had to mind her tongue when provoked. But now, with her royal pension, she was of independent means. Evidently William's workman Sprat was at hand when the row blew up, and no doubt he was both surprised and pleased when Caroline asked if she might leave William to lodge with him and his wife. They could do with the extra income, and Mrs. Sprat would be happy to clean and cook for Caroline.

Slough was tiny and Sprat lived only a short distance from William, and so it was a simple matter for William and Caroline to arrange to sweep of an evening. Sprat himself would also be in attendance, to raise and lower the tube of the reflector, and afterward he would be able to escort Caroline home. He could do the same if Caroline wished to sweep for comets when William was away. The note she made on July 30, 1798, is typical:

> My brother went with his family to Bath and Dawlish. I went daily
> to the Observatory and work-rooms to work, and returned home
> to my meals, and at night, except in fine weather [*sic*], I spent some
> hours on the roof, and was fetched home by Sprat.

Caroline often did her paperwork in her lodgings, but this proved inconvenient, for most of the books and records were with William: "Uncommonly harassed in consequence of the loss of time necessary for going backward and forward, and not having immediate access to each book or paper at the moment when wanted."

For some reason, after two years she left Sprat and moved into rooms with a "sober, industrious" tailor in the village, but this proved to be a mistake. Before the end of the year "The bailiffs took possession of my landlord's goods, and I found my property was not safe in my new habitation."

Caroline was paid by the king to be William's assistant in astronomy, but William regarded her simply as his assistant. Marriage had made him wealthy, and his enthusiasm for the rigors of nighttime observing was dwindling fast. He began to spend more and more time in Bath, where Alexander was prominent in the musical life and where William had other friends, not least William Watson, who had cemented their relationship by standing as godfather to John.

During his 1798 tour William began to think of how he could make worthwhile observations while he was away from Slough. To be portable, a reflector would obviously have to have a much shorter focal length than the 20 feet and 40 feet of his existing major instruments. This would call for mirrors that were thicker and more deeply ground than was normal with William's instruments, but he thought that by now he was equal to this challenge.

As a small-scale trial he experimented with a mirror of a respectable 7 inches in diameter but a focal length of a mere 35 inches. He "found that it would bear a magnifying power of 300 with great distinctness." Multiplying up, he found that a comparable reflector with a 24-inch mirror would have a focal length of 10 feet. Such a tube could—just—be transported by coach:

> Having an inclination to pass some time at Bath which I should not like to do without having an instrument of a large aperture I have it in view to make one of the above dimensions for the sake of ready transportation and that it may stand on a small place.[19]

And he added: "At all events the reduction of instruments to a small size is a very desirable object."

He rented a house on Sion Hill early in 1799 and, interestingly, subscribed to a local music library.[20] Indeed, a Bath paper announced that he planned to spend eight to ten months of the year there.[21] If this is true, he presumably intended to resign his position at court. The royal pension of £200 was now an irrelevance, but William was something of a snob; and "Slough, near Windsor" was a prestigious address from which to conduct his trade in telescopes.

The following year William gave up the Sion Hill house, which was some way out of town, in favor of one in Little Stanhope Street, around the corner from his old haunts in New King Street. He had his furniture moved there from Sion Hill, but his new house had been unoccupied for some time and much needed to be done to make it habitable. As so often with William's problems, Caroline was the solution. William simply instructed her to go to Bath and sort things out. She obeyed, as she always did, but her disapproval at being treated like a servant is evident.

She would have to make arrangements to cover her lengthy absence from Slough. She intended to lock away her furniture and books in the rooms she rented and to give William the key. "But on receiving information they would be seized along with my landlord's goods by bailiffs, I prepared the same night for their removal, and all was safely lodged in a garrett at [William's] by July 2 at night." It had been a close shave.

Caroline spent several months at Little Stanhope Street. But Alexander's home was equally in need of care and attention, for his housekeeper was elderly. And so when her brother left Bath, as was his custom each summer, and went to Slough to help William with the telescopes, Caroline set to work to "clean and repair his furniture, and making his habitation comfortable against his return." Then, at the end October, "I received notice that in about a fortnight I should be wanted at Slough," for before William set off for Bath, he wanted to show Caroline the work that needed doing in his absence. Then away he went, leaving her with the keys, "to make order and for despatching memorandums." Throughout her adult life she had been at William's beck and call, and now that she had a royal pension to be his assistant, William had less scruple than ever about beckoning and calling.

Caroline next went to lodge at Windsor with her nephew George Griesbach, one of the five sons of her sister Sophia who formed the backbone of Queen Charlotte's band. But Windsor was two miles from William's home, an unfortunate distance for daytime commuting and a major obstacle when William wanted to sweep for nebulae at night. And so, after four months, she moved to Chalvey, less than a mile from Slough, where she rented a small house from a woodcutter.

The Search for Nebulae Comes to a Halt

Caroline herself never made a worthwhile observation after she walked out on William. As for William, his observation of a convincing example of "true nebulosity" in November 1790 had settled the great theoretical dispute as to the nature of the milky patches in the sky, and in the three years before Caroline left him, his sweeps for nebulae had been all but abandoned. Now that she needed to commute if she was to help him sweep, his enthusiasm declined still further. But what were he and Caroline to do with the 428 nebulae they had collected since sending the second catalogue of one thousand to the printer? William had long ago promised in print to pursue the work "till the whole be completed," and this was a realistic possibility, even though most of the remaining areas of sky were around the North Pole and the 20-foot would have to be rotated north in order to reach them. In their prime, William and Caroline would have carried the project through in weeks. But they were not in their prime, William was bored with the work, and Caroline no longer on call the instant the skies cleared.[22]

But you can't just write off 428 nebulae, and the secret decision was made to raise the number to 500, publish them in a third catalogue, and quietly forget the regions still unswept. The work took forever, so casually did they go about it. In the whole of 1799 they swept on only four nights, and in 1800 on just a single night. At last the magic total was reached, and on June 29, 1802, Caroline sent to the Royal Society a catalogue of 500 nebulae and clusters.

William's Audience with Napoleon

A fortnight later William set off on a tour with a difference: a visit to Paris that would include an audience with Napoleon Bonaparte. He left Slough along with Mary, John, and Sophia Baldwin on July 13 and reached Dover on the seventeenth. The sea was "boisterous" and it was not until the twentieth that they could attempt the crossing to Calais. Despite the wait, they encountered a high wind, and Mary "was extremely frightened and very ill" and took refuge below. William preferred to lie down on deck "and did not stir, tho' a great deal of water dashed over me."[23] Arrived in France, they stayed in Boulogne at the Hotel Britannique, which despite its name was "très mauvaise," and progressed to Paris, with accommodation en route that varied from "très mauvaise" to "très bon."

On the twenty-fifth William was taken to the Paris Observatory, which owned a 22-foot reflector with a mirror just under 2 feet in diameter.[24] It was one of the very few telescopes that could match William's 20-foot for size, and so of course he had to see it. Not surprisingly, it did not pass muster as far as William was concerned: "I believe it impossible that it can have a fair chance for shewing objects as it ought to do."[25]

On the July 28 he dined with the great physicist and astronomer Simon de La Place, whose nebular hypothesis of the solar system chimed with William's vision of the large-scale universe, and on the thirtieth the two men breakfasted together. "His Lady received company abed; which to those who are not used to it appears very remarkable." There followed a succession of breakfasts, dinners, and visits to the famous sights, punctuated at regular intervals with meals *chez* La Place.

The climax came on Sunday August 8. The day began with breakfast with Madame La Place, and then in the early evening William dined with the Minister of the Interior, to take his leave. After dinner the minister took William, La Place, and the Anglo-American physicist Count Rumford to Malmaison, "the pallas of the first Consul." Madame Bonaparte took them round the gardens, and then the minister introduced William and Rumford to Bonaparte himself. William's account presents the First Consul as an intelligent and courteous man of wide interests. "He now led us to a room where after a short time spent in conversing he seated himself in a chair, and politely desired me to sit down." By this stage in his life William was expert in courtly customs, and noticing that no one else sat down, merely bowed his thanks and remained standing.

> The first Consul then asked a few questions relating to Astronomy and the construction of the heavens to which I made such answers as seemed to give him great satisfaction. He also addressed himself to M. La Place on the same subject, and held a considerable argument with him in which he differed from that eminent mathematician. The difference was occasioned by an exclamation of the first Consul's, who asked in a tone of exclamation or admiration (when we were speaking of the extent of the sidereal heavens) "and Who is the author of all this!" Monsieur Dela Place wished to show that a chain of natural causes would account for the construction & preservation of the wonderful system, this the first Consul rather opposed. Much may be said on the subject, by joining the arguments of both we shall be led to "Nature and nature's God."

The conversation then turned to, of all things, the breeding of horses in England. The evening had seen an encounter that William would remember with pride for the rest of his days.

He arrived home on August 25, to be greeted with the unwelcome news that Caroline had checked their catalogue of 500 nebulae and found the total to be only 497. They must go out and find three more; and this they did. In fact, in the last week of September they swept on three nights, quite like the old days, and collected ten nebulae—three to bring the catalogue up to the required number, and seven left over for any future catalogue. But although a couple of months of intensive work might well have completed the promised survey of the entire sky visible from Slough, William no longer had the stomach for the fray; and so they quietly added the seven, plus three more that Caroline found on going back through her records, to bring the present catalogue up to 510. The title with its mention of 500 was left unchanged, to foster the illusion that the unfinished work was still in progress.

Double Stars Reexamined

At the turn of the century, with the sweeps for nebulae all but abandoned, William had begun to reflect once more on the double stars that he had collected so assiduously long ago as an amateur in Bath. John Michell had argued that they were too numerous to arise from chance alignments and that most of them must comprise two stars that were companions in space, bound together by gravity.

In the interim, gravity had come to play a central a role in William's thinking about the evolution of the cosmos, and he was now inclined to agree with Michell; and so, in the theoretical introduction to the final catalogue of 500 nebulae, his second section discussed "the union of two stars, that are formed together into one system, by the laws of attraction." These two stars would move in ellipses around their common centre of gravity. But was there any evidence of this actually happening?

One of the most famous of double stars is Castor, which had been recognized as a double by G. D. Cassini as long ago as 1678. William had measured the angle of the line joining the two components in 1779, and he had remeasured the angle three times in the 1790s and then, more purposefully, from March 1800 onward. By a lucky chance he mentioned what he was doing to Nevil Maskelyne, who recalled that his predecessor as Astronomer Royal, James Bradley, had told him that in 1759 the line joining the components of Castor was parallel to the line joining Castor to its twin, Pollux. The methodical Maskelyne had made a written note of this information, and he managed to locate the note among his papers. William now had evidence of the change of angle of the two components stretching over nearly half a century. In a paper published in 1803 he estimated the period of revolution of the components to be "342 years and two months," astonishingly close to the modern value of 350 years.

For the other double stars that he remeasured, his information was more fragmentary and related to changes that had taken place over a mere two decades or so. Confirmation that the components of these "binary" stars are indeed bound together by the inverse-square law of Newtonian attraction (rather than by some other attractive force) therefore had to await systematic and careful observations by future astronomers. Only in 1827, after William's death, did the French astronomer Félix Savary publish measurements that put the matter beyond doubt, and John Herschel then altered the inscription over his father's tomb to include mention of "the vast gyrations of double stars."[26]

Difficult Times

In March 1803 Caroline had moved yet again, this time to rooms in Mary's old home at Upton, which was now let to a tenant. It was half a mile across the fields to William's property, a pleasant enough walk in daytime, but hazardous at night, especially if there was snow on the ground. On one occasion she injured her ankle and was laid up for a fortnight, but the Upton house was near enough to William's home for work to be sent over to her.

If the skies cleared and Caroline knew William wished to observe, she would rap on the window of the home of a local boy, saying (in her Hanoverian accent) "Please will you take me to my Broder," and he would get a lantern and escort her.

The years of observing in the intense cold and damp of Windsor had undermined William's health, and as he moved into his late sixties, his bouts of illness became ever more frequent. Caroline records the melancholy story of one sickness after another. In January 1807, "My Brother returned with a violent cough, added to a nervous headache." On February 26, 1808, William was "so ill that I was not allowed to see him, and till March 8 his life was despaired of, and by Mar. 10th I was permitted to see him, but only for two or three minutes, for he is not allowed to speak."

Caroline herself was one of those people who think the common cold beneath them and who therefore always suffer from influenza. When she still had thirty-four years to live, she wrote: "I, for my part, felt I should never be anything else but an invalid for life, but which I very carefully kept to myself, as I wished to be useful to my brother as long as I possibly could." But in 1808 she did indeed have a nasty scare. On October 2, "I was very ill and had Dr. Pope to attend me." Payments to Pope occur regularly in Mary Herschel's accounts, but Caroline was less easily satisfied. A week later, "dismissed Pope and went to Dr. Phips." This, however, was a serious error, for Phips decided she was going blind and had her kept for a fortnight in a darkened room, practicing the skills she was going to need when her sight failed. By November 20 Phips had realized his mistake. He "pronounced me out of danger for becoming blind, which he ought to have done much sooner, or rather not to have put me unnecessarily under such dreadful apprehensions."

Caroline revered William, but he was a somewhat distant god and she never felt able to confide in him. Her need for a confidant had been resolved in the most unexpected manner. A highlight of her Hanoverian childhood had been the weeks when she had been allowed to study needlework, a girl from a poor home tolerated in a class for the wealthy. One day in 1802, when she was visiting Windsor Castle, she was introduced to Mrs. Charlotte Beckedorff, who had recently been appointed Keeper of the Robes to Queen Charlotte. To Caroline's amazement Mrs. Beckedorff greeted her as a fellow pupil from that class of so long ago. The two became intimate friends, and it was a friendship that would last into extreme old age.

The wariness with which Caroline treated William did not apply to Dietrich, her sometime baby brother, and they spent a happy fortnight together when Dietrich visited Slough in the summer of 1806. He returned two years later, but this time in distressing circumstances. Hanover had become a pawn in the power struggle between Napoleon and the King of Prussia, and conditions in the city were so bad that Dietrich decided he had no option but to go to England as a migrant worker and send his earnings home to his family. Caroline later reflected on his stay and what it had meant for her:

> From the hour of Dietrich's arrival in England till that of his departure, which was not till nearly four years after, I had not a day's respite from accumulated trouble and anxiety, for he came ruined in health, spirit, and fortune, and according to the old Hanoverian custom, I was the only one from whom all domestic comforts were expected. I hope I have acquitted myself to everybody's satisfaction, for I never neglected my eldest brother's business, and the time I bestowed on Dietrich was taken entirely from my sleep or from what is generally allowed for meals, which were mostly taken running, or sometimes forgotten entirely.

When Dietrich arrived in 1808, Caroline was still living at Upton. Although she was within walking distance of William's home, the trek across the fields in all weathers to dance attendance on Dietrich must have been unwelcome, and in June 1810 she moved to the Annexe attached to the Crown Inn, a mere stone's throw from William. Caroline did all the redecorating herself, William taking her to Windsor to choose the paper for the sitting room. Yet although she was so close to her brothers she was lonely. The last three months of 1812, for example, she spent "mostly in solitude at home, except when I was wanted to assist my brother at night or in his library."

Caroline lived at the Annexe for four years. William had a gardener, Cock, and Mrs. Cock was paid a guinea a year to look after William's hens. She was happy to cook and clean for Caroline as well. But then Cock fell seriously ill—indeed it was clear that he would never work again—and Mrs. Cock could not leave him to attend to Caroline. In addition the inn wanted to expand into the Annexe. And so poor Caroline had to move yet again, "to a small cottage in Slough, at a considerable distance from my Brother." Once again Caroline's visits to William's home required planning. When Mary was absent, or the family was off on holiday, she would go over and spend the night there. Otherwise she would do her "calculating and copying" in her own cottage.

Not surprisingly she had unhappy memories of these endless upheavals. In old age she wrote to John's wife Margaret to say that her last quarter-century in England was spent "amongst beings of whom I was afraid." "I was obliged to change my habitation no less than 7 times, which was always attended with useless expenses, and what was still more precious, *loss of time.*" Yet she could not bring herself to eat humble pie and effect a reconciliation.

Decline of the Great Reflector

William's observation in November 1790 of a star surrounded with a halo of "true nebulosity" was a pivotal moment in the history of astronomy. But it was a disaster for the 40-foot, whose primary purpose had been to decide whether such nebulosity existed. Thereafter William would seldom find use for it in his astronomical research.[27]

Indeed, even before then it had been rarely used. After observing Saturn a handful of times in August and September 1789, on October 20 William tried using the 40-foot for sweeping after the manner of the 20-foot, but the outcome was disappointing. "The speculum [mirror] will not work well tonight. As I have no proper cover for it yet there has of late been a great condensation of moisture upon it in the day time, which must have injured the face very considerably." He swept again on December 2, and then on May 27, before abandoning the program. The 40-foot had a much-reduced field of view compared to the 20-foot, and to sweep the entire heavens would have taken centuries.

He now considered confining his sweeping with the 40-foot to the zone near the ecliptic, the path of the Sun across the sky. The known planets were never far from the ecliptic, and perhaps William would strike it lucky a second time and find another planet unknown to science—that would silence the critics of the great reflector! But after just four sweeps spread over eighteen months, from September 1791 to April 1793, this program too was abandoned. Indeed, his heart was never in it—for meanwhile, between February 1792 and March 1795, he was also sweeping the ecliptic with his 7-foot, the modest-sized instrument that had brought him triumph with Uranus.

Caroline, whose work as amanuensis was always beyond reproach, later assembled a file of the observations made with the 40-foot, and this must have been intended as a brief for the defense, for she did nothing similar for any of William's other telescopes. In an attempt after William's death to convince his son that the effort put into the 40-foot had not been in vain, she wrote him that "many [observations] must have been lost, being noted only either on slates or loose papers. . . . Owing to my not being, as formerly, the last nor the first at the desk (generally retiring as soon as the mirror was covered), the memorandums were often mislaid or effaced before I had an opportunity of booking them." The idea of Caroline's going to bed if she might thereby risk precious observations is preposterous, and that she should take blame on herself in this way is a sign of how defensive she and William were over the monster.

It is a token of the widespread disillusionment with the great reflector that in 1807 one of William's closest friends, Patrick Wilson, the man with whom he regularly met to talk about matters scientific, wrote to him:

> I don't know if as yet you have met with De La Lande's *History of Astronomy* for the year 1806. . . . There is a Paragraph, concerning you and the 40 Feet Telescope, evidently calculated to impress the belief of the *total Failure* of your noble Instrument.[28]

Wilson misrepresents what Lalande in fact has to say—the Paris astronomer had merely repeated what William himself had written about the

Figure 22. The reflector of 2-foot aperture but only 10-foot focal length, which William constructed in 1799 as a manageable instrument for his declining years. In 1814 he sold it to Lucien Bonaparte, younger brother of Napoleon. RAS W.5/5, no. 12, courtesy of the Royal Astronomical Society.

centuries it would take to sweep the heavens with the 40-foot—but clearly such opinions were in the air.

There was good reason for this. The 40-foot was cumbersome from the start compared to William's trusty 20-foot, but from the turn of the century it faced even stiffer competition, when William completed a 10-foot reflector with mirrors 24 inches in diameter (figure 22). As we have seen, one motive for such an instrument was the possibility of its being transported from Slough to Bath when William wished to spend time there. Indeed, the mirror was actually polished in Bath and may even have been cast there. Of course a 24-inch mirror had only a quarter the surface area of those of the 40-foot, but there were huge compensations. "The apparatus is excellent. I can get any object in less than half a minute." His appetite for observing was whetted once more. "I want much to look over the ecliptic again," he wrote to Caroline in August 1801, "to see[k] for another planet, and to find whether Piazzi's star is a real planet, and if so, whether it be furnished with satellites. For this reason my 10 feet shall have all possible distinctness." The search of the ecliptic began on September 1 and continued to the end of the year; but with no success.

The 10-foot was an ideal instrument for an elderly observer. William would now need a powerful incentive to summon the two workmen (and Caroline) out of bed, and maneuver the unwieldy 40-foot around the sky. Such incentives were few and far between. After his fourth and last sweep of the ecliptic on April 9, 1793, it seems he did not use the monster again until 1798. Thereafter he used it on average less than one night a year. His comment on July 29, 1813, is typical: "The mirror is so much tarnished that the image of Saturn was very imperfect."

Defects in the mirrors had been part of the problem. Although the first mirror had been too thin to maintain its shape when tilted in the tube, it took a good polish, and William was reluctant to abandon it entirely. Indeed, he went on polishing it until 1797. But Caroline was later to write to her nephew John lamenting the time wasted on this mirror, claiming that Alexander "would more than once have destroyed it secretly if I had not persuaded him against it."

William had chosen a different alloy for the second mirror so that it would keep its shape better when in use, but in consequence it tarnished more rapidly. Repolishing it was a major operation, for the mirror weighed nearly a ton. It had to be hoisted out of the tube, trundled on a carriage to the polisher, polished for a number of days by men recruited for the purpose (and even, in 1793, a horse), trundled back, and then reinstalled in the tube. He preferred to do this when the Bath music season was over and Alexander was free to help, but both men were getting on in years. Handling a ton of metal with a makeshift crane was dangerous work. Caroline vividly remembered the 1807 repolishing:

In taking the forty-foot mirror out of the tube, the beam to which the tackle is fixed broke in the middle, but fortunately not before it was nearly lowered into its carriage, &c., &c. Both my brothers had a narrow escape of being crushed to death.

And so the sad saga continued. Sometimes as much as four years would elapse between one polishing and the next, sometimes as little as one. William's observations were equally capricious. In May 1805 he examined the figure of Saturn, and then five whole years passed before he made another observation, this time of Uranus. But this did not stop him claiming his £200 a year for running costs, which he would continue to do with such persistence that in 1820 King George IV finally bowed to the inevitable and combined William's pension of £200 and expenses of £200 into a pension of £400.[29]

Not surprisingly, when the instrument had been left to rot for years on end in a fixed position, it did not respond well when summoned back to duty. Caroline explained the problems:

For instance, the P[olar] D[istance] clock disordered by not having been used for some time; the timepiece not having been regulated, nor *every one* of the out-door motions wanting oiling or cleaning; . . . and, in general, the first night the instrument is used after it has been left at rest for some time, it cannot be expected that all should go on without interruption or ease without a good mechanical workman had spent best part of the day looking over the motions.

William last observed with the 40-foot—though briefly—in August 1814, when he was seventy-five. He tried to view Saturn, but "the mirror is extremely tarnished," and with that the great reflector ended its hugely disappointing career as a scientific instrument. If Caroline is correct, his final attempt at polishing had taken place a few weeks before, in June, and the comment she then makes reveals why William had persisted in his losing battle to keep the telescope in commission:

My brother, being about this time engaged with re-polishing the forty-foot mirror, it required some time to restore order in his rooms before any strangers could be shown into them, and I again was assisting him to prepare for the reception of the Emperor Alexander and the Duchess of Oldenburg, &c.

For when, back in 1785, William had first applied to the king for money to build the instrument, he had flattered the royal vanity by declaring himself to be one "who ardently wishes to promote . . . the glory of the patron who supports him." George was rightly seen as the outstanding patron of astronomy, and indeed of science, in contemporary Europe: "seen" in the literal sense by guests at Windsor Castle, who after dinner were sent up the road to Slough to view for themselves the wonder of the age that the popular press compared to the Colossus of Rhodes and the Porcelain Tower of Nankin. And it was indeed a wonder. As Oliver Wendell Holmes later wrote:

> It was a mighty bewilderment of slanted masts, spars and ladders and ropes, from the midst of which a vast tube, looking as if it might be a piece of ordnance such as the revolted angels battered the walls of Heaven with, according to Milton, lifted its mighty muzzle defiantly to the sky.[30]

Within months of the telescope's completion the king's "madness" made him personally unfit to visit the Herschel home except on the rarest of occasions, but Queen Charlotte continued to entertain guests at the castle, and the stream of visitors to Slough was remorseless. In 1818, three decades after the monster saw first light, and with its constructor in his eightieth year, visitors to the observatory included: Princess Elizabeth and the Prince of Hesse Homburg with a count and two barons (April); the Prince and Princess Schaumburg von der Lippe (June); the Archduke Michael of Russia "with a numerous attendance" (July); Princess Sophia of Gloucester, the Archbishop of Canterbury, and several lords and ladies (August); and the Ertz Herzog Maximilian of Austria (October). Even though the telescope had long been abandoned as an astronomical research tool—if indeed it can be said ever to have fulfilled such a function—its role in the entertainment of visitors to the castle was undiminished, and the court astronomer had little option but to struggle, year after year, to maintain at least the semblance of an operational instrument.

The polishing of the mirror was now beyond him; but because almost nobody was ever allowed actually to look through the telescope (smaller ones were at the disposal of visitors), this was not immediately obvious, and callers could at least marvel at the structure as a whole. And so, in July 1818, William attempted once more to restore the mounting; "but the great heat he was exposed to by directing the workmen who are repairing the woodwork, &c., &c., of the 40 ft is surely too much for him." The work continued into August, "but he was all the time too ill."

The telescope had become a millstone around William's neck. "But we have all had the grief," Caroline later wrote, "to see how every nerve of the dear man had been unstrung by over-exertion; and that a farther attempt at leaving the work complete became impossible." Eventually in private he had to admit defeat: "The woodwork is fast decaying and cannot be effectually mended, and . . . I cannot recommend the 40 feet to be kept up."

> The difficulty of repolishing its mirror, which is tarnished, and preserving or restoring its figure when lost, is so great that if a larger mirror than a 20 ft. should ever be wanting, I am of opinion that one of 25 ft. with a mirror of 2 feet in diameter, such as I have made [for the King of Spain, plate 8] and which acted uncommonly well, should be a step between the 20 and 40 feet Instruments.

Yet he could not bring himself to make public admission of failure. For this Caroline paid a price:

> But when all hopes for the return of vigour and strength necessary for resuming the unfinished task [of polishing] was gone, all cheerfulness and spirits had also forsaken him, and his temper was changed from the sweetest almost to a pettish one; and for that reason I was obliged to refrain from troubling him with any questions, though ever so necessary, for fear of irritating or fatiguing him.

And yet we know that a triumph had been within its grasp. In 1810–11 William had been preparing to reflect in print on the observations of a lifetime and to commit to paper his grand conception of the construction and evolution of the cosmos. Until now he had rarely looked at a nebula through the 40-foot, but he now chose to view a handful from Messier's list and for three of them he used the great mirror. In the mid-1840s the Irishman William Parsons, who had become third Earl of Rosse on the death of his father in 1841, was to do much the same with his "Leviathan of Parsonstown." Rosse mistakenly thought he had resolved the Orion Nebula into stars, and many astronomers would overhastily take this as proof that all nebulae were star clusters, with unfortunate consequences. But then, when Rosse examined M51, he saw that it had a beautiful spiral form, and he made magnificent sketches of it that bear comparison with modern photographs. Today, when astronomers look back on what Rosse achieved, they forget about the spurious and misleading "resolutions" and instead remember him as the discoverer of the first spiral nebulae. Four spiral nebulae were among the nebulae that William now examined, but these he observed with lesser instruments. With the 40-foot he viewed two globular clusters, M15 and M72, and the Orion Nebula. But if William had happened instead to pick on M51, whose dramatic appearance is reflected in its modern name of the Whirlpool Nebula, it is very possible that he would have been able to see the spiral form. If so, the 40-foot would not be remembered today as an unmitigated disaster but as a triumphant success. In observational astronomy the line between success and failure can be thin.

The 40-foot had begun life with one musical performance in the great tube as it lay on the ground, and it was to end with another. In 1823 John told his aunt, "The forty-foot is no longer capable of being used, but I shall suffer it to stand as a monument." But as the woodwork deteriorated the structure became dangerous, and at Christmas 1839 John resolved to lower the tube to the ground, seal the various bits of apparatus inside, and have it painted every three or four years as a precaution against rot. But first the family had to take their leave in due fashion, and so on New Year's Eve they gathered in the tube to sing a light-hearted requiem composed by John:

> In the old telescope's tube we sit,
> And the shades of the past around us flit;
> His requiem sing we, with shout and with din,
> While the old year goes out and the new one comes in.

> *Chorus of Youths and Virgins:*
> Merrily, merrily, let us all sing
> And make this old telescope rattle and ring.[31]

And so on for eight verses. Caroline, now in distant Hanover, imagined that "none but from above were listening to, and joining their kindred in the chorus below!"

The end of William's career as an observer had come in 1814, when William was seventy-five years of age. That year he made his last observations with the 20-foot and the 40-foot, and he arranged to sell to Lucien Bonaparte the large 10-foot he had made for his own use. But he had a wealth of observations stored away, the fruit of endless night-hours spent in the cold and damp, when he had used his 20-foot to examine faint celestial objects that were visible to his eyes alone. Many of these objects were distant in space, and therefore in time, for the speed of light is finite: "a telescope with a power of penetrating into space . . . has also, as it might be called, a power of penetrating into time past." William estimated that he had seen light that had been a staggering two million years on its journey, and he realized that the object would still be visible to him even if the source of the light had been annihilated just under two million years ago. He shared this extraordinary insight with the poet Thomas Campbell, who afterward told a friend: "I really and unfeignedly felt at the moment as if I had been conversing with a supernatural intelligence."

On his deathbed in 1822, William would repeatedly ask for assurance that his irreplaceable observations were safe under lock and key, for they were the foundation for the four great papers on the construction of the heavens that he had published in *Philosophical Transactions* between 1811 and 1818. He imagined widely scattered particles of light attracting each other and slowly forming ever more condensed clouds, out of which in time stars would be born (figure 12). These stars in turn would attract each other to form clusters that at first were scattered but became ever more condensed as time passed, leading to what we would term "gravitational collapse," after which the whole process would start over again.

Only a being with a lifespan of endless ages could hope to witness this process actually taking place, but it comes to much the same thing if we humans systematically assemble examples of nebulae and clusters in a sequence of what William calls articles, classified according to their age: young, middle-aged, old.

[I]t will be found that those contained in one article, are so closely allied to those in the next, that there is perhaps not so much difference between them, if I may use the comparison, as there would be in an annual description of the human figure, were it given from the birth of a child till he comes to be a man in his prime.

Here we have the origins of the evolutionary cosmos of modern astronomy. This was not the end of the clockwork universe of Newton, but it was the beginning of the end.

John Takes Up the Challenge

In 1809, meanwhile, John had entered St. John's College, Cambridge.[32] Early in 1813 he graduated B.A. as Senior Wrangler, the prestigious first place among those awarded top honors in mathematics. That April he successfully competed for a college fellowship; although he would seldom be resident in the college, he was to hold his fellowship until 1829, when he disqualified himself by his marriage. Then, in May, after a little lobbying by his father, he was elected a Fellow of the Royal Society.

After his triumph in the mathematical examinations in Cambridge, John the polymath became fascinated by chemistry, and he spent his vacations pursuing experiments at Slough in a laboratory built in the garden at a safe distance from the house. Many different careers were now open to him, for he had the problem that he excelled in anything he put his mind to. His close friend, James Grahame (who one day would introduce John to his future wife), had recently left Cambridge for his native Scotland, where he was studying law—and enjoying it hugely. John was minded to follow his example. But it would be courteous to go through the motions of consulting his father, and so early in November 1813 he wrote to William. Had he felt seriously in need of advice he would surely have made the sixty-mile journey from Cambridge to Slough, but he did not. This was a mistake.

William was no enthusiastic churchgoer, but he had a profound belief in a God who had created the universe and who had populated it to the full with intelligent beings. He had as a matter of principle never discussed religion with his son, and John had grown up to become a firm believer in religion "established by nature" but a skeptic in the Anglican religion "as established by law."[33] Too many divines of his acquaintance treated Holy Orders not as the opportunity to spread the Christian Gospel, but as the passport to a life of comfort and even luxury.

Unfortunately this was exactly what William urged upon his son.[34] If he entered the Church, he could be sure of ample leisure for intellectual pursuits, and provided he preached morally improving sermons, he need not trouble himself too much over niceties of theology.

John was shocked at this suggestion that he take an ecclesiastical sinecure: "I cannot help regarding the source of church emolument with an evil eye." William in turn was outraged: "The miserable tendency of such a sentiment, the injustice and the arrogance it expresses, are beyond my comprehension." A churchman urging his congregation to love their neighbors was surely preferable to the barrister who spends half his time advocating what he knows to be a tissue of lies.

John was mortified. He had, he now realized, sent his father "a very wrongheaded and most foolish letter." Mary retrieved the offending missive and returned it to its author. Father and son sent each other peace offerings, John went home for Christmas, normal relations were restored, and in January John embarked on his legal studies in London with his parents' grudging acquiescence.

There he became friendly with William Hyde Wollaston, whose lectures in chemistry revived his interest in the subject, and James South, a surgeon who had married into money and retired from medicine to devote himself to astronomy and who was assembling an enviable collection of precision telescopes. Between them they had little difficulty in distracting John from the study of law. Chemistry began to take precedence in John's scheme of things—in August, for example, we find him telling his close friend Charles Babbage about his experiments with potassium. In March of the following year, 1815, John applied for the vacant chair in chemistry at Cambridge and lost by only a single vote. Writing to tell Babbage of the result, he assured him that "on the whole I believe it is better as it is—I am become half a mineralogist." This in turn became so serious an interest that in 1818 this extraordinarily versatile man would be urged to apply for the Cambridge professorship of geology.

By applying for the chemistry chair, John had signaled that he was prepared to abandon law and return to Cambridge, and in May 1815 one his former teachers at St. John's College wrote to offer him a college position as subtutor and examiner in mathematics. But the offer must have been made in hope rather than expectation, for this was a post at the very bottom rung of the academic ladder, teaching undergraduates in tiny groups. John declined.

John's health was never robust, and these competing stresses took their toll. In September he took himself off to Brighton for some weeks by the sea to recover, and there he began to see the attractions of the undemanding post at St. John's. No doubt to the astonishment of the college, he now decided to accept.

The post was indeed undemanding, and for this reason it proved hugely frustrating. As John told Babbage,

> You are pretty well aware what a job it must be to be set from 8 to 10 or 12 hours a day examining 60 or 70 blockheads, not one in ten of whom knows his right hand from his left, and not one in ten of whom knows anything but what is in the book. . . . In a word, I am grown fat, full and stupid. Pupillizing has done this—and I have not made one of my cubs understand what I would have them drive at.[35]

Nevertheless, the prospect of a life spent in his beloved Cambridge stretched out before him.

But it was not to be, for in the summer of 1816 his father's strength began to fail. William's life was drawing to a close, with so much still to be done. His positions and descriptions of nebulae needed to be checked and revised. His published catalogues were those of a natural historian and of little practical use to other observers. And he had never had the chance to examine the skies that were below the horizon at Windsor.

Although with Mary's money the Herschel family was of independent means, William had never attempted to deflect his son from following a conventional career. But now, with death staring him in the face, he saw things differently. Rarely did he go on one of his holiday tours without Mary, and while on tour he almost never failed to record in writing details of his day-to-day progress. But there is one exception: "A tour to Dawlish with John Herschel 1816 August 12." This time William simply notes, in the shaky hand of his old age, that they traveled via Bath and Exeter in each direction. There is no mention of the pouring rain between Bath and Dawlish or of the great difficulty they had in finding beds for the night. Even an accident they had when the axle of their coach broke goes unrecorded, for William's tour had a deadly serious purpose: when the moment seemed right, he spoke with John and pressed him to abandon Cambridge and become his apprentice in astronomy, so that William should not take his hard-won experience and insights to the grave.

A few days later John wrote to another Fellow of St. John's whose father had lately died:

> My father it is true is in no immediate danger—he is even better than he has been for these two months—but I cannot be insensible to the great alteration which has taken place in him.

John goes on to speak of William with the words of the Roman poet Horace: "We rarely find anyone who can say he has lived a happy life, and who, content with his life, can retire from the world like a satisfied guest."[36]

John's life teaching mathematics as a "hired hack" was far from satisfying, yet he relished the brilliant company he enjoyed in Cambridge, and it was only with the greatest reluctance that he could bring himself to accede to William's wishes. As he wrote to Babbage,

> I always used to abuse Cambridge as you well know with very little mercy or measure, but, upon my soul, now I am about to leave it, my heart dies within me. I am going, under my father's directions, to take up the series of his observations where he has left them (for he has now pretty well given over regularly observing) and continuing his scrutiny of the heavens with powerful telescopes.

John wrote finis to his past life that December, when he sold his law books for six pounds.

Back in Slough, father and son, now master and apprentice, agreed that the 40-foot was beyond redemption; but the 20-foot needed no more than an input of youthful vigor under the supervision of a trained hand. The mirrors however were in a poor state and so, in 1817, the new team of father and son had two new ones cast. The first John polished under the supervision of the aged William, the second he polished himself. John was now equipped with the telescope that had served William so well, in full working order.

Or so he thought. In December 1820 John was visited by South. Together they turned the 20-foot on the Moon and Saturn. John was mightily impressed: "I can now believe anything of the effect of reflectors of great aperture." But the woodwork was in poorer condition than they realized, and the telescope "took it into its head to fly from its center." There was nothing to be done but to dismantle the instrument, repair the damaged components, and rebuild it.

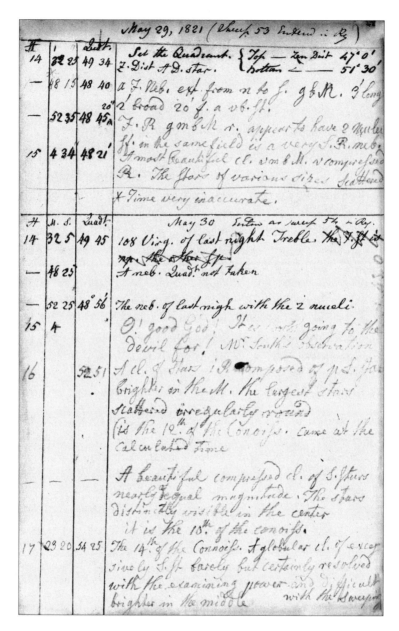

Figure 23. In 1821, nearly two decades after they ended their campaign of sweeping for nebulae and when William was well into his eighties, William and Caroline initiated John into the technique of sweeping. Caroline's pencil recorded the comments of James South, who was present, on seeing a nebula with two nuclei: "O! good God! It is worth going to the devil for!" RAS J.1/1, courtesy of the Royal Astronomical Society.

I am not sorry for this; it will afford my poor Father some occupation, which (though able to do very little) he stands much in need of, and is quite a new man when superintending some little repairs, &c.[37]

At first John regarded his 20-foot as a new instrument, but on reflection he preferred to see it as the last of William's, "and the clearness and precision of his directions during its execution, showed a mind unbroken by age, and still capable of turning all the resources of former experience to the best account."[38]

John now knew how to make a major telescope; but he still had to learn to sweep with it. And so, on May 29, 1821, a poignant scene was enacted at Slough. After darkness fell, the eighty-two-year-old William, summoning all his reserves of strength, helped his son to position the mirror in the tube. Caroline—now in her seventies—was at her desk again after an interval of two decades, ready to copy down the observer's shouted comments. Was William able to climb up and join John on the observing platform? Surely not; and yet the descriptions Caroline wrote down have an authentic ring about them.

They swept for just half an hour. Next evening they swept again, this time for no less than three hours. South had been invited and was with John on the observing platform, and Caroline duly copied down what he too had to say when shown a nebula with two nuclei: "O! Good God! It is worth going to the devil for!" (figure 23).

Now it was up to John; William could do no more. Caroline writes that "one day passed like another, except that I, from my daily calls, returned to my solitary and cheerless home with increased anxiety for each following day." She concentrated on the autobiography she was writing for Dietrich. Months passed, and William grew ever more frail. On August 15, 1822, he was visited by a grandson of the Bulmans with whom he had lodged in Leeds so long ago, who begged a token he might send to his father. Caroline was sent to the library to fetch one of William's papers, and a plate of the 40-foot reflector. Ten days later William was dead.

12

1822–1833
John's "Sacred Duty"

John had sacrificed his chosen career in Cambridge for what he would later describe as his "sacred duty" of completing his father's work. But around the time of William's death, John was often to be found in London. He was already a recognized member of the scientific establishment, deeply involved in the affairs of both the long-established Royal Society and the infant Astronomical Society of London, which he had helped to found and of which William had been the nominal first president. Another reason for his absences in London lay in James South's precision telescopes, which were ideal for the scrutiny of double stars, and the two men were enjoying an exhilarating collaboration in the very field where William had cut his astronomical teeth. But the peerless 20-foot stood idle in the garden at Slough; and soon John's conscience began to prick him.

William had demonstrated in equal measure the three skills that in those days made a great astronomer. As the builder of telescopes capable of reaching far out into space (and time past) he was unrivalled. As an observer he had himself surveyed almost the whole of the sky visible from Slough. And as a theorist he had boldly used these observations to argue that the universe was not clockwork in nature but evolutionary, not mechanical but biological.

But there were major limitations in the store of observations that William had bequeathed his successors. The most obvious of these was that William had never seen the skies below his horizon: his catalogues, be they of nebulae and clusters or of double stars or of the comparative brightness of stars, were plainly incomplete.

To remedy this was a task for the future. Of more immediate concern were the problems resulting from William's introduction into astronomy of the specimen collecting of the natural historian. His vast catalogues of nebulae and clusters, in particular, had been organized simply by class and then by date of discovery, and an arrangement of less use to the observer would be hard to imagine. How could one possibly determine whether a nebula or cluster that one encountered was already in William's lists? The three catalogues needed to be combined and then reorganized into a format that observers could actually use.

Not only that, but William and Caroline had worked in conditions that called for heroism, with William shouting out his descriptions and locations to a Caroline so cold that the ink sometimes froze in her inkwell. Had she heard him correctly? Had the reference stars been correctly identified? Had errors of transcription crept in before the data finally appeared in *Philosophical Transactions*? John was his father's heir in astronomy, and it was his duty to reexamine each of his father's nebulae, confirm (or correct) its location and description, and then prepare a unified catalogue that suited the needs of observers. But how was he to do this? It would be absurd for him to maneuver his 20-foot reflector toward the first nebula of the first class, track it down and subject it to due examination, and then rotate the instrument toward a completely different region of sky in search of the second of the class—even supposing that this region was currently visible. That would take forever.

In practice John could reexamine the two and a half thousand nebulae only by proceeding as William had done when discovering them: by directing the reflector to the south and making systematic "sweeps" of the heavens as they rotated overhead, until he had scrutinized the entire sky visible from Slough. And for this reexamination he needed to have the catalogues of nebulae merged and completely recast into a format that would allow him to prepare for each night's observing by making a list of the objects he might expect to encounter. But the one person who could be relied on to undertake this vast project and carry it through to a conclusion was now back in her native Hanover.

Caroline Returns to Hanover

Caroline had given years of anxious thought to what she should do when William's death finally put an end to her duties as his assistant.[1] Her pension of £50 for life had been confirmed by King George IV, and William's will provided for her to have an annuity of £100,[2] so she was financially secure; but should she remain in England or return to Hanover? The animosity Caroline had felt against Mary in the early years of William's marriage had long since disappeared, to be replaced by a somewhat shame-faced affection and respect. Mary was "a dear sister, for as such I now know you." But they had little in common beyond their love for William and John, and Caroline had chosen to live in rented accommodation for the last two decades rather than "eat humble pie" and ask to be accepted back into the cottage that was attached to the Herschel home where she had once held sway.

The talented and charming John was of course a joy to his childless aunt, but he was frequently absent from Slough, and in any case, in his current investigations of bright double stars he had no more need for an amanuensis than had William in Bath all those years ago.

England, in short, offered her little. Hanover by contrast offered a lot. She had the roseate memory of her youthful haunts that we all have (and she would become disillusioned by the reality, as we all do). Alexander, for so many years the regular summer visitor to Slough, had retired to his native city in 1816, after an accident had put an end to his career as a cellist (William funded his retirement by sending him £50 twice a year). Her confidant Mrs. Beckedorff had done the same soon after the death of Queen Charlotte in 1818. Not long after, Caroline had made up her mind to retire to Hanover:

> I sent in 1820 thirty pounds to be laid out for a fether bed for me when after a long dreaded melancholy event [the death of William] I should be obliged to seek consolation in the busom of Diterich's family, which after the description of his wife and daughters' characters, I thought to be the only place on earth where I could find rest.[3]

The following year Alexander died. This was sad—Caroline had long had a special affection for the brother whose personality had been so damaged by his apprentice-master—but the attraction of Hanover remained over-whelming: her sometime baby brother, Dietrich, was there with his wife and children and grandchildren, "noble-hearted and perfect beings," all eager to welcome her into the bosom of their family and give her the love she craved. Anxious to cement this embryonic relationship, shortly before William's death Caroline had sent Dietrich her life savings of £500 to help him in a family crisis.

When William at last passed away, the mechanics of winding up her affairs in the country that had been her home for half a century provided Caroline with a distraction from her grief. She had been impetuous in 1772 in entrusting her future to William and England, she had been impetuous in 1778 in rejecting the offer of a singing engagement that opened up the possibility of a career in music, and she had certainly been impetuous in flouncing out of the Herschel home in 1797 and going into rented accommodation. Now she was impetuous in quitting England for Hanover.

It was to prove a disastrous decision without a single redeeming feature, and one that was to blight the final quarter-century of her long life. "Oh!" she would one day write to John's wife, "why did I leave England!"[4]

John bore the brunt of her rushed exit, which took place when he still had his father's affairs to settle and his mother to console.

> My aunt though greatly distressed has borne this affliction with uncommon fortitude. She has resolved on leaving England immediately and going to reside with her family in Hanover, and the expectation of preparing for her journey has been of service in distracting her attention from dwelling on its cause.[5]

John understood very well the blunder Caroline was making, and he had already tried to dissuade her from giving the £500 to Dietrich. But the one argument that might have swayed her was the one argument he felt it would be improper for him to use: the help she could be to him in his future work. In her heart of hearts Caroline knew that when she opted for Hanover she was abandoning William's scientific heir, leaving him to complete her brother's work as best he could without her. Her conscience would trouble her for the rest of her life. But she did not expect to live long. Caroline was a healthy hypochondriac convinced that each month would be her last, and she was already in her seventies.

Her eagerness for Hanover verged on panic. She refused to wait until John had settled the most urgent matters and could see her safely there. As he wrote to Dietrich,

> My aunt's fixed determination to quit England before the winter will render it impossible for me to do what I had earnestly wished, viz. to see her in Hanover, as it will be impossible for me to arrange all my dear Father's affairs soon enough and I cannot yet leave my Mother who requires my presence and support.

It was at this point that Dietrich showed the depth of his affection for Caroline. Only a few weeks earlier his eldest daughter Anna had been widowed within a few days of giving birth to her ninth child. Traveling to her home to offer what help he could, Dietrich himself had become ill. Yet he made the difficult journey to England to fetch his sister.

Dietrich arrived at Slough on October 3, 1822, six weeks after William's death. From then on events moved fast. On the seventh Caroline took her leave of Princess Augusta and other Windsor friends. On the tenth she and Dietrich left for London. On the fourteenth Princess Matilda sent her carriage for Caroline so that they could spend the day together. Two days later, at Bedford Place, London, "all my friends were assembled," among them John and his mother "Lady Herschel," as Caroline continued to call her sister-in-law in accordance with the customs of the time. Caroline was asked to sign a receipt for the first half-year's annuity under William's will. She always felt awkward about accepting money she had not earned, especially if it smacked of a brotherly handout (and in the years ahead John would have to exercise patience and ingenuity in coaxing her to accept the money due to her, money that was small change as far as he was concerned). But after persuasion she agreed, and later admitted that it "has enabled me to furnish myself with many conveniences on my arrival here, of which otherwise I should have perhaps debarred myself"; in other words, the money had come in handy.

Caroline, John told a friend, was "well & in good spirits. She has exerted herself and made all her arrangements with extraordinary vigour." Caroline was in denial. Deep inside her she was already beginning to realize the enormity of the mistake she was committing. The feather bed, the money given to Dietrich—neither had involved an irrevocable commitment to Hanover. But she had declared her intention, and now she was carried along by the momentum she herself had generated. For her, England was a land of grief, of memories of good times now gone forever. Writing long afterward, Caroline recalled that "From all these sorrowing friends and connections I was obliged to take an everlasting leave," but the imagined obligation was entirely of her own making.

Before their departure Caroline and Dietrich lodged in an inn near the Tower of London, where they dealt with the customs formalities. John "came for a moment to us, and after his departure I saw no one I knew or who cared for me." The reality was beginning to dawn. "My fears for what was to come and regret for what I left behind were so stupifying that it made me almost insensible to all what was passing about me."

Next morning they boarded the steam packet. It had been half a century since the first voyage of Caroline's life, and the seas had not improved in the meantime. It took them forty-eight hours of misery to reach Rotterdam.

> At one time a spray conveyed a bucket-full of water into my bed, which was regarded as nothing in comparison to the evils with which I was surrounded. I was the most sick of all on board, and the poor old lady was pitied by all. . . . At Blackwall we lay still three hours, then we hobbled on to near Gravesend, and there lay in a high sea at anchor all night, whilst they were hatching and thumping to mend the vessel we were to go in [to be taken ashore]. At half past eleven I set foot on shore, where so many people were assembled to gaze on us that it set me a crying.

It was next day that the awful truth began to dawn. In the past Dietrich had been a confidant, a soul mate, whereas William had been a respected and loved, but nevertheless domineering, elder brother to whom she could never open her heart. But the Dietrich who was with her in the coach taking her to her new home in Hanover was not the Dietrich she had last known in Slough a decade ago. His earlier affection and generosity of spirit had been blighted by suffering. Bad enough that his daughter Anna had been recently widowed and had nine young children to support, but there was worse. Dr. Johann Richter, the talented physician married to Dietrich's second daughter, Dorothea, had become insane, able to practice medicine only in brief periods of lucidity. So great would his physical threat to his wife and four children become, that Dietrich would one day have to move them to a place of safety. And Dietrich himself was in indifferent health. As a result, he was a changed, even embittered man.

> But in the last hope of finding in Dietrich a brother to whom I might communicate all my thoughts of past, present, and future, I saw myself disappointed the very first day of our travelling on land. For let me touch on what topic I would, he maintained the contrary, which I soon saw was done merely because he would allow no one to know anything but himself.

What especially rankled with Caroline was Dietrich's complaint that their father had denied him a proper education. Isaac, "that excellent being," had been the one person in Caroline's life who had never ever failed her and in her eyes was beyond criticism—as indeed he seems in truth to have been. Dietrich had been no more than eleven years old when Isaac had died, too young to appreciate the heroic efforts their father had made to further the lad's welfare as his own health declined. Now in his late sixties, Dietrich "for ever murmured at having received too scanty an education, though he had the same schooling we all of us had before him."

But then, with their arrival in Hanover a week later, things took a turn for the better. Although Dietrich's wife Catharina was plagued by rheumatism, she took exercise and kept cheerful, "and her reception of me was truly gratifying." Caroline was to have an apartment in the same house; it had rooms several times the size of those she had been used to in Slough and was "furnished in the most elegant style." Before long, Dietrich's grandchildren were lined up to be presented to a Caroline who had never known a family with more than one youngster since she herself had been a child. The grandchildren, informed that this was Great-Aunt Caroline, were mystified: at four-foot something, Caroline looked to them to be anything but great.

And there was Mrs. Beckedorff, her one and only close friend outside of the family, who had sent to inquire after her within a couple of hours of her arrival. But Hanover itself was a great disappointment—of course. "It is quite a new world, peopled with new beings, to what I left it in 1772." And the streets that had once been large had shrunk over the half-century.

"An Extraordinary Monument in the Cause of Abstract Science"

But how was she to occupy her days? Never before had Caroline's life lacked purpose. On the contrary, whether it was cleaning forks in Hanover or training a choir in Bath or copying observations in Slough, she had seldom had a moment's respite. But a suitable challenge was at hand: within a year of her arrival in "*Horrible Hannover*,"[6] she undertook for John her last and arguably greatest contribution to the fulfillment of William's ambitions. It would be an accomplishment that the Scots physicist Sir David Brewster later described as "an extraordinary monument of the unextinguished ardour of a lady of seventy-five in the cause of abstract science."[7]

When William had come across a nebula, he had defined its position by reference to a nearby star, along the lines of "up by so much, left by so much." To reorganize the catalogues in the manner John required, the first step was to assemble these reference stars into a list suited to the needs of the observer engaged in sweeping: firstly, by the angular distance of the star from the North Pole (for all the objects seen in a given sweep would be at much the same North Polar Distance), and secondly, in the order in which they would present themselves as the heavens rotated overhead. At William's request, Caroline had embarked on this preliminary task in the summer of 1799. It was a long-term project that she pursued when she had nothing more urgent on hand, and it took her until 1818 to finish the job.

This hard work was now about to bear fruit. In August 1823 John wrote to tell her that his 20-foot reflector was "in fine order." He had finished for the time being with the nearby double stars that called for small precision telescopes, and he was about to turn his attention to his father's distant nebulae. Caroline replied at once, saying that she wished to live a little longer "that I might make you a more correct catalogue of the 2,500 nebulae," that is, one designed for the observer engaged in sweeping. Caroline already had the reference stars arranged in the required format; now all she needed to do was to use these star positions to assemble a matching list of nebulae.

It was a straightforward enough task, but mammoth in scale, calling for heroic perseverance and near-impeccable accuracy with figures. Fortunately John was able to visit her in the autumn of 1824 on the way home from one of his European tours, to discuss with her what would be for him the ideal format. John's impression of life in Hanover is revealing: "I found her very comfortably situated in her brother's family, and with no cause to regret her change of country." His visit, of course, coincided with the few months when she had purpose in her life.

Caroline's handwritten folio volume reached John in April 1825. Now in the Royal Society Library in London, it runs to 104 pages of numbers. As John was later to write, "I learned fully to appreciate the skill, diligence, and accuracy which that indefatigable lady brought to bear on a task which only the most boundless devotion could have induced her to undertake, and enabled her to accomplish." His response was immediate and heart-warming: "These curious objects [the nebulae] I shall now take into my especial charge—nobody else can see them."[8] Two years later, when his sweeps were in full swing, he wrote to his aunt:

> I find your Catalogue most useful. I always draw out from it a regu-lar working *list* for the night's sweep, and by that means have often been able to take as many as thirty or forty nebulae in a sweep.[9]

Remembering her two decades of night watches in partnership with William, Caroline's conscience troubled her as she imagined John strug-gling simultaneously to make the observations and to record them:

> I . . . am only sorry that I cannot recall the health, eyesight, and *vigor* I was blessed with twenty or thirty years ago; for nothing else is wanting (and that is all) for my coming by the first steamboat to offer you the same assistance (when sweeping) as, by your father's instructions, I had been enabled to afford him.[10]

Perhaps declining health would not have permitted her to work all night at a desk within shouting distance of the reflector. But she could easily have spared John the resulting daytime paperwork; and she knew it.

Back in London, the Fellows of the Astronomical Society were well aware of the crucial role Caroline's catalogue was playing in John's work. Only men could be admitted to fellowship, but this restriction did not ap-ply to the society's medals, and in 1828 the Council unanimously resolved

> That a Gold Medal of this Society be given to Miss Caroline Her-schel for her recent reduction, to January, 1800, of the Nebulae dis-covered by her illustrious brother, which may be considered as the completion of a series of exertions probably unparalleled either in magnitude or importance in the annals of astronomical labour.[11]

Because John was then the president, it was fitting that the announce-ment be made by South as the vice-president; and Caroline's less-than-enthusiastic comment on his tribute to her was an early sign of her increas-ing obsession with the defense of William's reputation: "I could say a great deal about the *clumsy speech* of the V.P. Whoever says *too much of me* says *too little of your father!*" Perhaps it is as well that the discussion in the Royal Society, no less, as to whether she deserved its prestigious Royal Medal fell afoul of the narrowly conceived time constraints that then applied to the period when the work to be honored had been carried out.[12]

But occasionally Caroline allowed herself a modest pat on the back. As she wrote to John:

> Of Alex and me can only be said that we were but tools and did as well as we could; but your Father was obliged to turn us first into those tools with which we could work for him; but if too much is said in one place [in praise of me], let it pass; I have perhaps deserved it in another by perseverance and exertions beyond female strength! Well done![13]

John's Reexamination of William's Nebulae and Clusters

John had embarked on reobserving his father's nebulae and clusters in the late summer of 1825, a few weeks after Caroline's volume reached him. His mechanic John Stone systematically raised and lowered the tube of the 20-foot during each sweep, but in the absence of Caroline John had to act as his own amanuensis. His procedures were subject to

> modifications rendered necessary by the loss of my aunt, Miss CAR-OLINE HERSCHEL's personal assistance, on whom the task of reading and registering the Polar distance and Right Ascension of objects, writing down the remarks and descriptions, warning the observer of expected stars, and finally reducing and calculating the whole, used invariably to devolve. Unsupported by such aid, I am under necessity of recording the observations myself.[14]

And so at last Caroline was paid the public tribute that William had never thought called for.

Fortunately for John the absence of Caroline was not the crippling handicap it would have been for his father, for Caroline's catalogue warned John in advance of when it would be safe to interrupt observing and make written notes. But it was to be a long campaign, extended over some eight years. We glimpse just how demanding it was from a letter John wrote to a friend in February 1828:

> Every hour is precious, from the circumstance of the great mass of nebulae lying in the 11, 12 and 13 hours of R.A. [the celestial counterpart of longitude], and which therefore must be observed in the spring or not at all. A pellucid sky—the total absence of moonshine

and twilight—and *nebulae to look at*, are conditions which co-incide, on the average, not 20 nights in the year, and the sacrifice of a night . . . is therefore a very serious evil to me, regarding as I do the completion of this work not as a matter of choice or taste, but a sacred duty which I cannot postpone to any consideration.[15]

At last the job was done. In 1833 *Philosophical Transactions* published John's catalogue of 2,307 nebulae and clusters, of which 525 were new.[16] Over 700 of William's objects were missing, clear evidence of the need for the revision. And now the entire list was systematically ordered, so that an observer coming across a nebula could easily check whether or not it had already been listed by John.

But there remained the greater challenge, that of extending William's work to the southern skies.

13

1833–1848
"The Completion
of My Father's Work"

John Examines the Southern Skies

The major limitation of John's 1833 catalogue was stated in its title: the observations were "made at Slough." It was time for John (plate 12) to extend William's work to the skies below the horizon at Slough. He had not been able to consider leaving for the Southern Hemisphere as long as his aged mother depended on him for emotional support, but Mary had died in January 1832. Not only that, but—except for legacies totaling no more than £2,000[1]—she had left her entire estate to John. This would easily allow him to meet the expenses of transporting his growing family to the Cape of Good Hope and housing them while he carried out one of the great campaigns in observational astronomy.

In June 1832, John (now Sir John) visited Caroline:

> I found my aunt wonderfully well and very nicely and comfortably lodged, and we have since been on the full trot. She runs about the town with me and skips up her two flights of stairs as light and fresh at least as *some folks* I could name who are not a fourth part of her age. . . . In the morning till eleven or twelve she is dull and weary; but as the day advances she gains life, and is quite 'fresh and funny' at ten or eleven p.m., and sings old rhymes, nay, even dances! to the great delight of all who see her.

He discussed his plans with her, and the old warhorse scented battle: "Ja! If I was thirty or forty years younger, and could go too? In Gottes Namen!"[2]

On November 13, 1833, the *Mountstuart Elphinstone* set sail from Portsmouth with the Herschel entourage on board: John, his wife Margaret, three children, a nurse, John Stone, the 20-foot, a refractor that John had purchased from South for precision observations, Caroline's large sweeper (for a first reconnaissance of the southern skies), and a mountain of luggage. They arrived two months later and were soon established in an eighteen-room house south of Cape Town.[3]

In the clear South African climate it took John just four years to extend to the skies below the Slough horizon every aspect of his father's campaigns. His sweeps resulted in another 1,708 nebulae and clusters of stars. He found 2,103 double stars, and he counted nearly 70,000 stars distributed over some 3,000 areas.[4] He sketched,[5] he mapped, he observed lunar eclipses, he devised a gadget to measure the brightness of stars, he drew pictures of scenery and flowers with a camera lucida, he went exploring, he even helped devise a new educational system for the colony. And when they set sail for home in March 1838 his family had increased from three to six.

When John was at the Cape, Caroline remembered that William (unaware, of course, of the existence of interstellar dust) thought that in Scorpio there was a space totally empty of stars. "As soon as your instrument is erected I wish you would see if there was not something remarkable in the lower part of the Scorpion to be found, for I remember your father returned several nights and years to the same spot, but could not satisfy himself about the uncommon appearance of that part of the heavens." John did as instructed and found the region full of beautiful clusters of stars.

This was not the right answer. "It is not *clusters of stars* I want you to discover in the body of the Scorpion (or thereabout), for that does not answer my expectation, remembering having once heard your father, after a long awful silence, exclaim, 'Hier ist wahrhaftig ein Loch im Himmel!' [Here indeed is a hole in the heavens!]" The diplomatic John lost no time in reassuring her that in Scorpio there were *both* clusters *and* spaces devoid of stars.

Back in 1793 the *Bath Chronicle* had told its readers that William expected soon to give an account of the lunar inhabitants.[6] A far more elaborate series of claims—this time a deliberate hoax intended to boost circulation—was published by an enterprising contributor to the New York *Sun*. It took the form of an article supposedly taken from an Edinburgh journal, and it was published in six installments at the end of August 1835. John, it appeared, had taken to the Cape a giant refracting telescope whose principal lens was 24 feet in diameter and weighed nearly seven tons, and with it he had identified on the Moon woods and meadows, hills and valleys, and even living organisms. He had, for example, "classified not less than thirty-eight species of forest trees." Mammals identified by John included a beaver, "which carries its young in its arms like a human being, and moves with an easy gliding motion."[7]

These ludicrous announcements created a sensation and were widely believed. It was of course months before John became aware of what was happening, and months more before his denials were published. By then his "moon men" were the talk of coffeehouses not only in America but in Europe, and in 1836 a German publishing house even put out a complete book on the subject.[8]

Caroline Meets William James

When John arrived home from the Cape, it was to great acclaim, and the hereditary title of baronet was conferred on him on the occasion of Queen Victoria's coronation in June 1838. So famous was he that a letter from Germany with his name and the simple address "London" was duly delivered.

He lost no time in setting out for Hanover, accompanied by his five-year-old son William James. It was a time for celebration. Caroline, now eighty-eight, was delighted, and not least to have living proof that her brother's name was continued in another generation of the family. But she was unused to children, and four of her own siblings had died young. She was sure that the lad would come to harm if she took her eyes off him for a moment. And if she gave him something to eat, it would poison him. "I rather suffered him to hunger than would let him eat anything hurtful; indeed I would not let him eat anything at all without his papa was present." But the lad's great-aunt had unconventional entertainment to offer William James, as an amused observer records: "Well! what do you say of such a person being able to put her foot behind her back and scratch her ear, in imitation of a dog, with it, in one of her merry moods."

John and Caroline both knew that this would be the last time they would meet, and Caroline prepared to make a speech of final farewell. She had much to say. Her brother could now rest in peace, his life's work—and hers—brought to a triumphant conclusion by John. But when the hour came for her peroration, she found that John and William James had already fled the town, unable to face up to the pain of parting.

Caroline was devastated. "You came to take Tea with me and soon left me for ever!" she wrote indignantly. "Without taking leave!" Never could she go through such a traumatic experience again, and in 1842 she told John's wife Margaret (whom she never met but for whom she had great affection): "I would not wish on any account to see either my nephew, or you, my dear niece, again *in this world, for I could not bear the pain of parting once more;* but I trust I shall find and know you in the next."

Caroline's Last Years

Caroline (plate 13) was an honored figure in Hanover, and not only in Hanover. In 1832 she was awarded a medal by the King of Denmark for her earlier discoveries of comets, but the tribute was unwelcome: it "provokes me beyond all endurance, for it is no use to me." Much more welcome were her election in 1835 as Honorary Fellow of what had become the Royal Astronomical Society and the award three years later of membership of the Royal Irish Academy. She would have both these honors recorded on her tomb. But she became obsessed with little things. She repeatedly debated what should be done with her few possessions, and she repeatedly assured John that he must not worry about her funeral expenses because she had made adequate provision.

Like the miser of legend counting his coins, Caroline several times reckoned how much she had given Dietrich and his family over the years, right down to the last *gutengroschen.* But whereas the miser got pleasure from the results of his tally, Caroline's calculations made her ever more resentful as she calculated "how much I have thrown away on beings to whom I was under no obligation of *any* kind; and among whom there is not one who would sometimes sacrifice an hour to cheer me in a long winter's evening." The family to whom she had given her life's savings begrudged every minute spent with the lonely old lady. "But instead of experiencing the least attention from those who are in Hanover or from the absent Nieces when they are in Town I hardly come in for a hasty call just before they are on their way to dinner."

This was an ominous sign of what was to come. Gradually the poison entered into Caroline's soul, at first slowly, and then, after Dietrich's death in 1827, with increasing pace. Dietrich himself she would remember for boasting of his wealth "so as to demand homage from everyone for being a capitalist . . . and a brother of Wm Herschel," to whose legacy of £2,000 Dietrich in fact owed his financial standing. Dietrich's wife ceased to be the cheerful woman who had received Caroline in a manner that was "truly gratifying." Instead she was transformed in Caroline's imagination into "a short corpulent woman upwards of 60, dressed like a girl of 20 without cap, her brown hair mixed with gray plated and the temples covered in huge artificial curls I almost shuddered back from her embrace." Caroline would have us believe that the widow Catharina now revealed herself to be a woman of "avarice," who "never comes near me without she has some design on my purse."

In similar vein Caroline no longer remembered the apartment prepared for her on first arrival as "furnished in the most elegant style." It became in her embittered memory a "room which had been washed in the forenoon, no fire, there I dropt on a sopha my eyes fixed on the wet boards without a carpet which it was my first care to buy and one for Mme H. by way of giving no offence by showing a desire of having anything apart." And with the aging Mme. Beckedorff seldom to be seen, Caroline had no one in whom to confide, and so the poison festered within her: "of the events of the last 10 years I have spent here, I can only say that they have been a perfect tissue of disgusting vexations doubly painful to bear because I could not communicate my complaints to anyone; because they were against my immediate connections."

Caroline was also becoming obsessed with defending William's reputation. His achievement in deep-sky observations had been made possible by the great reflectors he had built. But the 40-foot with its 48-inch mirrors had been a failure, and both William and John's 20-foot reflectors had mirrors just 18 inches in diameter, impressive but clearly not at the limit of the available technology. In the late 1830s the future Lord Rosse took up the baton, and on his estate in what was then King's County in Ireland set out to surpass William's achievement in the construction of great reflectors. In 1839 he completed a reflector with a mirror made of segments with a 36-inch diameter. The following year he cast an integral mirror of the same size, and then began planning a monster reflector with mirrors no less than 6 feet in diameter.[9]

John, who understood better than anyone the monumental challenge Rosse had set himself, was full of admiration; and when the "Leviathan of Parsonstown" was at last completed in 1845 (plate 14), he would declare it to be "an achievement of such magnitude . . . that I want words to ex-press my admiration for it."[10] Innocently supposing that Caroline would be equally delighted that others were building on the foundations that William had laid, in 1841 John sent her an article describing what Rosse had managed to do so far. But far from being delighted, Caroline was out-raged. Rosse's "great telescope, which *shall* beat Sir William Herschel's all to nothing"—the idea was so absurd that she spent the rest of the day in laughter. In 1844 she wrote to John saying that "They talk of nothing here at the Clubs but of the great Mirror and the great Man who made it. I have but one answer for all; which is *Der Kerl ist ein Narr!*"—the guy is a jerk![11]

The same year she embarked on the writing of a second account of her early years.[12] She had written the first in the 1820s, at first for Dietrich and then, when Dietrich died, for John. It ended—tantalizingly for us—with William's marriage, and so tells us nothing of the difficult years that fol-lowed. This second version was also for John, whom his ninety-year-old aunt asks to "excuse the style and the spelling, &c, &c, on account of my eyesight being so uncertain"; but she assures him that "my memory is as good as ever," as indeed it was. A year later, she sent him thirty-one pages of fair copy, written as ever in her firm and legible hand, telling her life story to the time of her arrival in Bath in 1772. Margaret wrote to beg her to "*go on* with your memoir until you leave England and take up your residence in Hanover." She did her best to respond. Her ambition was to cover the period from 1772 to the present, but in October 1842 she com-mented, "What a Hotchs Pot! How shall I get through it?" She struggled on, more and more slowly. In April 1844 she told John she was continuing to devote her energies to "writing the *Life & Adventures of Miss Caroline Hersche[l]*," and in August 1845, when she was ninety-five, John wrote to thank her for the latest installment, which was again written in the clearest of hands and shows a memory undimmed with age. It ends with Dietrich's abortive break for freedom in 1777 and his arrival at Bath. It is an aston-ishing achievement for a lady in her nineties, writing in a language that was not native to her and which she had not spoken for two decades. Of her efforts to integrate herself into Bath society seventy years before, she remarks:

> Writing the above brings to my recollection the answer I gave once to a Lady, when reproving me for being my own Trumpeter, by say-ing "how can I help it? I cannot afford to keep one!"

John urged her to persevere—"If it is only at the rate of 3 or 4 lines a day it will be an amusing occupation. . . . You have no idea how it interests us and Marg‍ᵗ is quite in raptures when she gets a few fresh pages of it"— but her strength was at last beginning to fail.

In 1846 the King of Prussia awarded her the Gold Medal for Science, "In recognition of the valuable services rendered to Astronomy by you as the fellow-worker of your immortal brother, Sir William Herschel." The same year the Astronomer Royal, the formidable George Biddel Airy, called to see her. Like all her visitors, Airy was astonished at the vigor and intelligence of a woman well into her nineties; as always, Caroline was full of grumbles.

> She was not up when we called at 11 but was up at 2. She complained of not being well, but seemed to us to be extremely well, and possessing powers of body and mind which I could never expect to see in a person of such an age. She spoke partly in German and partly in English, of course we lost a little, especially as her articulation is slightly defective, but generally we got on very well.[13]

On her ninety-seventh birthday she was visited by the Crown Prince and Princess, who brought her the gift of a velvet armchair. She entertained them for two hours and sang them a catch that William had written.

"You Will Then Have in Your Hands the Completion of My Father's Work"

Meanwhile John had been hard at work, preparing his Cape observations for publication. The data in his notebooks embodied the extension to the southern skies of William's (and Caroline's) observing campaigns, but as long as these data remained unpublished, they were of no use to the astronomical community. Unfortunately it was not a matter simply of seeing pages of numbers through the press. The data as recorded by John were expressed in coordinates local to his observatory near Cape Town and were affected by factors such as atmospheric refraction. They must be converted into universal coordinates and purged of the unwanted complications: they must be "reduced." This was a routine task in arithmetic, and many observers would have delegated it to a reliable assistant; but not John. He slogged his way through them one by one, a task that took years.

Caroline Herschel

Figure 24. Engraving showing Caroline at the age of 97. Miss Beckedorff, daughter of Caroline's close friend Mme. Beckedorff and herself a support of Caroline in her final years, was dissatisfied with it. "The artist has, I believe, imitated the style of the old German school of Albert Durer, resembling more a 'woodcut' than a print, nor does it do justice to her fine old countenance. Yet it is extremely like in feature, expression, and deportment, her eyes have taken the languid expression more from fatigue occasioned by her *sitting* for the picture whilst she is used generally to recline on her sofa, and I see them very frequently sparkle with all their former animation."

And meanwhile Caroline was moving into advanced old age. She had been eighty-eight when John paid her his final visit, before the reductions were even started, and as he worked steadily away, so the months remaining to her slipped past. When she was ninety-four she wrote John's wife a letter in which she expressed her fears that she would die without seeing the tangible proof that William's work had been completed and the completion made available to observers worldwide.[14] At long last, in July 1847, when Caroline (figure 24) was well into her ninety-eighth year, John was able to send her the sumptuous volume of his Cape observations. It had been the best part of a decade in the making, and it is an astonishing production, worthy of the extraordinary four-year-long campaign of which it was the record. In 1816 John had made a solemn promise to his aged father: his "sacred duty" was to accomplish everything in astronomy that William would have done, had not advancing years prevented him. Now, three decades later, his promise had been discharged in exemplary fashion. The volume he sent Caroline bore the proud title (figure 25) *Results of Astronomical Observations Made . . . at the Cape of Good Hope, Being a Completion of a Telescopic Survey of the Whole of the Visible Heavens, Commenced in 1825*. In fulfilling his promise, John had become the first observer in history—and he would be the last—to examine every part of the entire celestial sphere with a major telescope.

When the book reaches you, he proudly told Caroline in a separate letter, "You will then have in your hands the completion of my father's work."[15] She was no longer able to write, so we have no record of the tears of joy she no doubt shed. She had lived beyond the age of ninety-seven in the hope of seeing this day, and now, like Simeon with the baby Jesus, she could utter her Nunc Dimittis. A few months later, on 9 January 1848, she died.

Her funeral took place on the eighteenth. As a mark of respect the King of Hanover, and the Crown Prince and Princess, sent their coaches to follow the hearse; and by order of the Crown Princess, her coffin was adorned with palm branches. Inside her coffin, at her request, was placed a lock of her brother William's hair and an almanac that had been used by her father.

She had long before purchased the plot of ground where her beloved father had been buried back in 1767. Twenty-two years later, her mother Anna had been buried in the same grave. Because there was no room in the

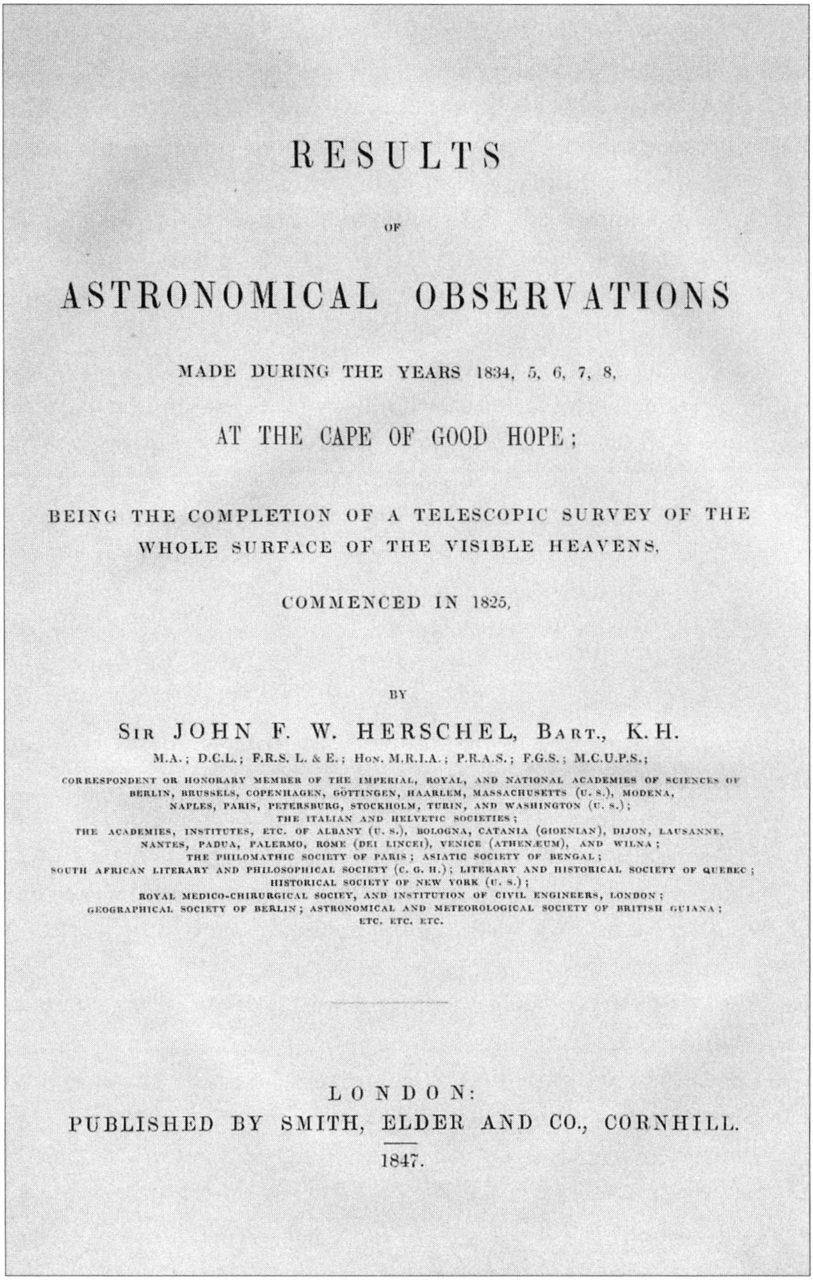

RESULTS

OF

ASTRONOMICAL OBSERVATIONS

MADE DURING THE YEARS 1834, 5, 6, 7, 8,

AT THE CAPE OF GOOD HOPE;

BEING THE COMPLETION OF A TELESCOPIC SURVEY OF THE
WHOLE SURFACE OF THE VISIBLE HEAVENS,

COMMENCED IN 1825,

BY

Sir JOHN F. W. HERSCHEL, Bart., K.H.

M.A.; D.C.L.; F.R.S. L. & E.; Hon. M.R.I.A.; P.R.A.S.; F.G.S.; M.C.U.P.S.;

CORRESPONDENT OR HONORARY MEMBER OF THE IMPERIAL, ROYAL, AND NATIONAL ACADEMIES OF SCIENCES OF
BERLIN, BRUSSELS, COPENHAGEN, GÖTTINGEN, HAARLEM, MASSACHUSETTS (U. S.), MODENA,
NAPLES, PARIS, PETERSBURG, STOCKHOLM, TURIN, AND WASHINGTON (U. S.);
THE ITALIAN AND HELVETIC SOCIETIES;
THE ACADEMIES, INSTITUTES, ETC. OF ALBANY (U. S.), BOLOGNA, CATANIA (GIOENIAN), DIJON, LAUSANNE,
NANTES, PADUA, PALERMO, ROME (DEI LINCEI), VENICE (ATHENÆUM), AND WILNA;
THE PHILOMATHIC SOCIETY OF PARIS; ASIATIC SOCIETY OF BENGAL;
SOUTH AFRICAN LITERARY AND PHILOSOPHICAL SOCIETY (C. G. H.); LITERARY AND HISTORICAL SOCIETY OF QUEBEC;
HISTORICAL SOCIETY OF NEW YORK (U. S.);
ROYAL MEDICO-CHIRURGICAL SOCIEY, AND INSTITUTION OF CIVIL ENGINEERS, LONDON;
GEOGRAPHICAL SOCIETY OF BERLIN; ASTRONOMICAL AND METEOROLOGICAL SOCIETY OF BRITISH GUIANA;
ETC. ETC. ETC.

LONDON:
PUBLISHED BY SMITH, ELDER AND CO., CORNHILL.

1847.

Figure 25. The title page of the book with which John Herschel completed his father's work, by extending William's observational campaigns to the skies below the horizon of Windsor.

earth for a third coffin, in 1825 Caroline had a vault built over the same spot, and this can be seen today in the Gartenkirchhof on the Marienstrasse. On the upper slab is the inscription that she herself composed. Her first draft, dated July 26, 1844, had been brief, but it included reference to her membership of the Royal Irish Academy and of the Astronomical Society of London (its original name), of which she was so proud: "Beneath this Stone are deposited the remains of Caroline Herschel M.R.I.A. & A.S.L. who died this . . . month 184. aged &c. In or near this place Her Father Isaac Herschel was buried March 25, 1767. Aged 60 years, 2 Months and 17 days."[16] It is a German translation of her undated second draft that the visitor finds inscribed on her tomb (plate 15):

> Here rest the earthly remains of Caroline Herschel. Born at Hanover 16th March 1750. Died . . .
>
> The gaze of the deceased while here below, was turned towards the starry heavens; her own discovery of comets and her share in the immortal labours of her brother William Herschel will testify hereof to future generations. The Royal Irish Academy and the Royal Astronomical Society of London counted her among their members. At the age of years months days she fell asleep in calm and cheerful possession of all her powers of mind, following to a better world her father Isaac Herschel who after attaining the age of 60 years 2 months 17 days was buried at this place 25th March 1767.[17]

She tempted fate by declaring in advance that she had died in possession of all her powers of mind, yet for once fate was kind to her.

Her inscription announces the presence of her father in the same tomb; but of her mother, whose body lies to this day above that of Isaac and below that of Caroline, she makes no mention. In a final act of revenge for Anna's attempt to imprison her at home as a lifelong drudge, and to deny her her role in the greatest family enterprise astronomy has ever seen, she condemned her mother to an unmarked grave.

Abbreviations

BL   British Library

Burney   *Diary & Letters of Madame D'Arblay* [Fanny Burney], ed. Char-
              lotte Barrett (6 vol., London, 1905)

CHA   *Caroline Herschel's Autobiographies*, ed. Michael Hoskin (Cam-
              bridge, 2003)

Chronicle   *The Herschel Chronicle: The Life-story of William Herschel and
              his Sister Caroline Herschel*, ed. Constance A. Lubbock (Cam-
              bridge, 1933)

Crowe   M. J. Crowe et al., eds., *A Calendar of the Correspondence of Sir
              John Herschel* (Cambridge, 1998)

Hanover   Michael Hoskin, *The Herschels of Hanover* (Cambridge, 2007)

HFA   Herschel Family Archives (private collection)

JHA   *Journal for the History of Astronomy*

Memoir   *Memoir and Correspondence of Caroline Herschel*, by Mrs. John
              Herschel, 2nd ed. (London, 1879)

"Memorandums" "Memorandums from which an historical account of my life
              may be drawn," by William Herschel, Royal Astronomical So-
              ciety, The Herschel Archive, W.7/8

Partnership *The Herschel Partnership: As Viewed by Caroline*, by Michael
              Hoskin (Cambridge, 2003)

PT	*Philosophical Transactions of the Royal Society*
RAS	Royal Astronomical Society, The Herschel Archive
WH	William Herschel

Citations of the form *Chronicle*, chap. 16, page 19 are to the original typescript of *Chronicle*, in the possession of the William Herschel Society, Bath.

The manuscripts of William, Caroline and John Herschel that are held by the Royal Astronomical Society are listed in J. A. Bennett, "Catalogue of Archives and Manuscripts of the Royal Astronomical Society," *Memoirs of the Royal Astronomical Society*, 85 (1978). William Herschel carefully preserved incoming letters, and these are in RAS W.1/13; his book of outgoing letters is RAS W.1/1. All are individually listed by Bennett. Most letters between members of the family are still in the HFA, except those from Caroline in Hanover in her old age, which are in the British Library, Egerton 3761 and 3762.

Notes

Prologue ▪ August 1772: The Partnership Convenes

1. This account of Caroline's arrival in England is based mainly on her own recollections, *CHA*, 117–18.

2. Caroline to John Herschel, August 21, 1838, BL, Egerton 3762.

3. Private and Personal Acts 1793, c. 38. The handwritten copy signed by the Clerk of the Parliament was sold at the Sotheby's Herschel auction in 1958, Lot 473. So English did William become that he wrote to German friends (e.g., Schroeter) in English.

4. "As it was my lot to be the Ashenbröthle of the Family (being the only girl) I could never find time for improving myself." Caroline to Margaret Herschel, September 28, 1838, BL, Egerton 3762.

1 ▪ 1707–1773: A Musician's Odyssey

1. For information on Isaac and Anna (and on each of their children), see the biographical essays in *Hanover*. For Caroline's early life in Hanover, see *CHA*.

2. Jürgen Hamel, "Ein Beitrag zur Familiengeschichte von Friedrich Wilhelm Herschel," Gauss-Gesellschaft E. V. Göttingen, *Mitteilungen*, 26 (1989), 99–103.

3. J. B. Sidgwick, *William Herschel: Explorer of the Heavens* (London, 1953), 21, quoting John Herschel.

4. On this episode see the account in *Partnership*, 13–15.

5. Michael Hoskin, "Was William Herschel a Deserter?," *JHA*, 35 (2004), 356–58.

6. William's compositions are today being played once more. The best list of the surviving manuscripts is in Ronald Lessens, *William Herschel: Musicien Astronome* (Vannes, 2004), Annexe VI.

7. Robert Smith, *Harmonics, or The Philosophy of Musical Sounds* (Cambridge, 1749). William's copy was later sold at the Sotheby's Herschel auction in 1958, Lot 446.

8. Robert Smith, *A Compleat System of Opticks*, 2 vols. (Cambridge, 1738).

9. "Memorandums," 13–14.

10. *Chronicle*, 18.

11. Angus Armitage, *William Herschel* (London, 1962), 21, quoting John Herschel.

12. WH to Jacob, July 10, 1761, *Chronicle*, chap. 2, p. 19.

13. WH to Jacob, January 22, 1762, *Chronicle*, chap. 3, p. 24.

14. English translation in the HFA.

15. "Therefore I was engaged in the Army when I was about 15 years of age, where I remained till my 19th year, when I quitted the Service and settled in England." WH to G. C. Lichtenberg, February 15, 1783, RAS W.1/1. "The known encouragement given to Music in England determined me to try my fortune abroad & accordingly ab[t] the year 1759 I came to settle in this country." WH to Charles Hutton, November 1784, quoted from Owen Gingerich, "William Herschel's 1784 Autobiography," *Harvard Library Bulletin*, 32(1) (Winter, 1984), 73–82, p. 77.

16. For Alexander's apprenticeship see the biographical essay in *Hanover*.

17. From "Memoirs of George Ludolph Jacob Griesbach," in Griesbach family possession.

18. *Leedes Intelligencer*, no. 548, October 9, 1764.

19. Quoted without reference by J. B. Sidgwick, *William Herschel: Explorer of the Heavens* (London, 1953), 30.

20. *Chronicle*, 37.

21. *Leedes Intelligencer*, no. 651, September 2, 1766.

22. On the Octagon Chapel and its organ, see A. J. Turner, *Science and Music in Eighteenth Century Bath* (Bath, 1977), 31–33.

23. For an account of Bath at this time, see Roy Porter, "William Herschel, Bath, and the Philosophical Society," in G. E. Hunt, ed., *Uranus and the Outer Planets* (Cambridge, 1982), 23–34.

24. For further details of William's musical activities in Bath, see Michael Hoskin, "Vocations in Conflict," *History of Science*, 41 (2003), 315–33.

25. Edward Rack, "A Disultory Journal of Events &c at Bath," Municipal Reference Library, Bath, MS 1111, 5.

26. Ian Woodfield, *The Celebrated Quarrel Between Thomas Linley (Senior) and William Herschel: An Episode in the Musical Life of 18th Century Bath* (Bath, 1977).

27. John Marsh, memoirs, vol. 9, 754, Cambridge University Library, Add. MS 7757.

28. For information on Caroline's time in Bath, see *Partnership*, chap. 2.

2 ▪ 1773–1778: Vocations in Conflict

1. For further information on the activities of William, Caroline, and Alexander in Bath, see *Partnership*, chap. 2.

2. RAS W.5/12.1.

3. Ibid. Caroline is more explicit: "all his rubish of patterns, Tools, Hones for grinding, polishers, unfinished mirrors &c. &c., but all for small Gregorian, none above 2 or 3 inches dia[mete]r." *CHA*, 52.

4. WH to G. Newenham, October 30, 1781, RAS W.1/1.

5. WH to G. C. Lichtenberg, July 18, 1785, RAS W.1/1.

6. *CHA*, 124–26.

7. John Bernard, *Retrospections of the Stage*, ed. W. B. Bernard, 2 vols. (London, 1830), vol. 2, 60–61.

8. "For the observatory at Oxford," September 3, 1774, MS Radcliffe 29, Museum of the History of Science, Oxford. The price was £4.14.6.

9. H. C. King, *The History of the Telescope* (London, 1955), chap. 5.

10. James Ferguson, *Astronomy Explained upon Sir Isaac Newton's Principles* (London, 1st ed. 1756), section 56.

11. RAS W.3/1.4, 7–10, p. 8.

3 ▪ 1779–1781: An Enthusiasm Shared

1. W. J. Williams and D. M. Stoddart, *Bath: Some Encounters with Science* (Bath, 1974), 68.

2. Roy Porter, "William Herschel, Bath, and the Philosophical Society," in G. E. Hunt, ed., *Uranus and the Outer Planets* (Cambridge, 1982), 23–34, p. 30.

3. Galileo was not the first to propose this method, as is often supposed. See Harald Siebert, "The Early Search for Stellar Parallax: Galileo, Castelli, and Ramponi," *JHA*, 36 (2005), 251–71.

4. John Michell, "An Enquiry into the Probable Parallax, and Magnitude of the Fixed Stars . . . ," *PT*, 57 (1767), 234–64.

5. WH to A. Aubert, January 9, 1782, RAS W.1/1.

6. A. Aubert to WH, January 22, 1782, RAS W.1/13.A.8.

7. Royal Society, Letters and Papers 1741–1806, VII.192.

8. N. Maskelyne to WH, April 23, 1781, RAS W.1/13.M.15.

9. RAS W.5/12.1, 48–58.

4 ▪ 1781–1782: Royal Patronage

1. A. F. O. Alexander, *The Planet Uranus* (London, 1965), 37.

2. Royal Society MS EC/1781/19. He was nominated by W. Watson Sr., W. Watson Jr., J. Smeaton, C. Blagden, J. Lloyd, H. Stebbing, F. Wollaston, and S. Hemming.

3. *Chronicle*, 112.

4. *Bonner and Middleton's Bristol Journal*, March 30, 1782.

5. He proposed to show the king the celebrated Gamma Virginis, Gamma Leonis (whose double nature he had discovered only on February 11, 1782), Pi Bootis (September 20, 1779), 54 Leonis (February 21, 1781), Castor (Alpha Geminorum, April 8, 1778, independently discovered by Christian Mayer), Alpha Herculis ("rather obscure and difficult," August 29, 1779), Beta Cygni (September 12, 1779), and Gamma Andromedae (August 25, 1779), RAS W.2/1.4, 12. Except for Gamma Leonis, all had been identified as doubles by earlier observers.

6. *Chronicle*, 115.

7. WH to Alexander, June 10, 1782, HFA.

8. W. Watson to WH, June 12, 1782, RAS W.1/13.W.16.

9. WH to Caroline, July 3, 1782, HFA.

10. From "Memoirs of George Ludolph Jacob Griesbach," in Griesbach family possession.

5 ▪ 1782–1783: "Astronomer to his Majesty"

1. On the events of the following years as seen by Caroline, see *Partnership*, chap. 3, and *CHA*.

2. Advertisement in the *Reading Mercury*, May–June 1785.

3. As William asked to be styled in the certificate of membership of the Hollandish Society of Sciences. WH to Martin van Marum, June 22, 1791, RAS W.1/1.

4. WH, "On the Proper Motion of the Sun and Solar System . . . ," *PT*, 73 (1783), 247–83, reprinted with discussion of this and the ensuing papers in Michael A. Hoskin, *William Herschel and the Construction of the Heavens* (London, 1963), 43–59.

5. Arthur Eddington, "Herschel's Researches on the Structure of the Heavens," *Occasional Notes of the RAS*, 1 (1938–41), 27–32, p. 30.

6. The early history of variable stars is discussed in Michael Hoskin, "Novae and Variables from Tycho to Bullialdus," *Stellar Astronomy: Historical Studies* (Chalfont St Giles, 1982), 22–28.

7. For the work of Goodricke and Pigott, see Michael Hoskin, "Goodricke, Pigott and the Quest for Variable Stars," *JHA*, 10 (1979), 23–41.

8. E. C. Pickering, *Harvard Annals*, 23 (1890), 231.

9. T. Collinson to WH, December 5, 1794, RAS W.1/13.C.19.

10. For a full discussion of William's investigation into the spectra of stars and its background, see Michael Hoskin and David Dewhirst, "William Herschel and the Prehistory of Stellar Spectroscopy," *JHA*, 37 (2006), 393–403.

6 ▪ 1783–1785: The Construction of the Heavens

1. Scattered clusters are in fact components of our Galaxy, while globular clusters are external satellite systems. In his 1789 paper on the construction of the heavens, William goes to great lengths to invent ways in which scattered clusters might evolve into globulars, but he is far from convincing.

2. For further details of William's studies of nebulae, see Michael Hoskin, "William Herschel's Early Investigations of Nebulae: A Reassessment," *JHA*, 10 (1979), 165–76.

3. Michael Hoskin, "Caroline Herschel's Catalogue of Nebulae," *JHA*, 37 (2006), 251–55, and "Caroline Herschel: Assistant Astronomer or Astronomical Assistant," *History of Science*, 40 (2002), 425–44.

4. Royal Society MS 272.

5. B. Faujas Saint-Fond, *Voyage en Angleterre, en Écosse, et aux Îles Hébrides*, vol. 1 (Paris, 1797), 74–89; Translated as *Travels to England, Scotland, and the Hebrides*, vol. 1 (London, 1799), 63–77. A much simpler account of the visit is given in the manuscript by Faujas Saint-Fond transcribed by Audoin Dollfus, "Une visite chez William Herschel," *L'astronomie*, 101 (1987), 135–46.

6. John Smeaton to John Michell, November 4, 1785, RAS MS Radcliffe Hornsby 78.

7. *Astronomisches Jahrbuch für das Jahr 1788* (Berlin, 1785), 162–64, trans. in *Chronicle*, 138 (with error as to source).

8. "For instance, an equal scattering of the stars may be admitted in certain calculations; but when we examine the milky way, or the closely compressed clusters of stars, . . . this supposed equality of scattering must be given up." WH, "Astronomical Observations Relating to the Construction of the Heavens . . . ," *PT*, 101 (1811), 269–336, p. 270.

7 ▪ 1782–1790: "One of the Greatest Mechanics of his Day"

1. WH to C. Mayer, October 8, 1782, RAS W.1/1.

2. *Hanover*, 88.

3. It has been speculated that William's purpose in building large telescopes was to confirm his belief in intelligent life elsewhere in the universe; but he expressly told Faujas de Saint-Fond that high magnification was *not* his goal: "Les nébuleuses de M. Messier sont nébuleuses avec le télescope de 7 pieds et lorsqu'on les observe avec le télescope de 20 pieds [de Herschel], l'on voit claire-ment que les nébuleuses sont un amas d'étoiles. L'intention de M. Herschel, en faisant ce si grand télescope [de 40 pieds], n'a pas pour but l'agrandissement de l'objet, mais de se donner, à l'aide des plus grands miroirs, la plus grande quan-tité de lumières possibles." Audoin Dollfus, "Une visite chez William Herschel," *L'astronomie*, 101 (1987), 135–46, p. 141. In a letter of July 9, 1793, to an un-named correspondent (RAS W.1/1) William says, "The chief excellence of the instrument is its power of penetrating into space owing to the great quantity of light it collects."

4. For further information on the financing of the 40-foot, its early triumph, and its decline, see Michael Hoskin, "Herschel's 40ft Reflector: Funding and Functions," *JHA*, 34 (2003), 1–32.

5. Charlotte L. H. Papendiek, *Court and Private Life of the Time of Queen Charlotte* (London, 1887), vol. 1, 191.

6. *Chronicle*, 146.

7. Richard Bentley, *Some Stray Notes on Slough and Upton* (privately printed, 1892); *Chronicle*, 146.

8. *Chronicle*, 145.

9. See Slough History Online, www.sloughhistoryonline.org.uk, accessed January 1, 2010.

10. Maxwell Fraser, *The History of Slough* (Slough, 1973), 55–56.

11. Charles G. Harper, *The Bath Road: History, Fashion, & Frivolity on an Old Highway* (London, 1899), 8.

12. Ibid., 9.

13. The Newbury Society Web site, www.newbury-society.org.uk/timeline.htm, accessed January 1, 2010.

14. Burney, vol. 3, 18.

15. Letter of M-A. Pictet of Geneva, published in *Journal de Genève*, 1787, 45–62, p. 49, trans. in *Chronicle*, 158.

16. *The Diaries of Colonel the Hon. Robert Fulke Greville*, ed. F. M. Bladon (London, 1930), 96, referring to November 22, 1788.

17. *The Later Correspondence of George III*, ed. A. Aspinall, vol. 1 (Cambridge, 1962), letter 379.

18. *Memoir*, 308.

19. Burney, vol. 3, 148.

20. Caroline to John, October 27, 1830, BL, Egerton 3761.

21. E. G. Forbes, "The Pre-Discovery Observations of Uranus," in Garry Hunt, ed., *Uranus and the Outer Planets* (London, 1982), 67–80, p. 75. William himself discusses such possible observations of Uranus in his letter to J. Banks, December 27, 1816, RAS W.1/1. Galileo observed Neptune on December 28, 1612, and January 27, 1613. Charles T. Kowal and Stillman Drake, "Galileo's Observations of Neptune," *Nature*, 287(5780) (September 25, 1980), 311–13; and Stillman Drake and Charles T. Kowal, "Galileo's Sighting of Neptune," *Scientific American*, vol. 243(6) (1980), 52–59.

22. J. Banks to WH, November 17, 1789, RAS W.1/13.B.24.

8 ▪ 1786–1788: "Gold Can Glitter as Well as the Stars"

1. Burney, vol. 3, 322.

2. In the words of Fanny Burney's father, Charles, writing in 1797, Burney, vol. 5, 345.

3. Burney, vol. 4, 113. For further information on William's courtship of Mary and their marriage, see *Partnership*, 91–94.

4. Pitt's will is in the National Archives, Kew, England, Prob 11/1018.

5. Adee Baldwin's will is in the National Archives, Kew, England, Prob 11/945. John Herschel discusses the amount (and the fate) of Nathaniel Phillips's estate in papers on the history of his mother's family, HFA.

6. The National Archives, Kew, England, Prob 11/1100.

7. Her will is in the National Archives, Kew, England, Prob 11/1255. She left pecuniary legacies in excess of £2,300, and the rest was to be divided between Mary and John Herschel, and Mary's brother's family. John Herschel, in "Recovery of dates," HFA, writes: "She left by her will 5000 3% stock in joint names of my mother & myself. She lived at Walton. Her home there she and property at Greenwich & 'The Crown at Walton' left jointly to my Mo^r and M^r B^n" [Thomas Baldwin].

8. Charlotte L. H. Papendiek, *Court and Private Life of the Time of Queen Charlotte* (London, 1887), vol. 2, 147–48.

9. *Chronicle*, 177.

10. Mary Clark's will is in National Archives, Kew, England, Prob 11/1255. John Herschel, "Recovery of dates" (HFA) remarks on the amounts involved with some bemusement.

11. Elizabeth Baldwin's will is in National Archives, Kew, England, Prob 11/1314.

9 ▪ 1788–1798: "Noble and Worthy Priestess of the New Heavens"

1. For more on Caroline as an observer, see Michael Hoskin, "Caroline Herschel as Observer," *JHA*, 36 (2005), 373–406; and *Partnership*, chap. 3.

2. Nevil Maskelyne to Nathaniel Pigott, December 6, 1793, RAS Nathaniel Pigott archives.

3. *Memoir*, 92.

4. *Memoir*, 101.

10 ▪ 1788–1810: "The Most Celebrated of All the Astronomers of the Universe"

1. J. de Lalande to WH, April 26, 1787, RAS W.1/13.L.5, envelope.

2. For more on William's search for Uranian moons, see A. F. O. Alexander, *The Planet Uranus* (London, 1965), chaps. 3 and 4.

3. A. Aubert to WH, March 4, 1801, RAS W.1/13.A.33.

4. *Edinburgh Review*, Jan. 1803.

5. On the history of Bode's Law see Michael Martin Nieto, *The Titius-Bode Law of Planetary Distances: Its History and Theory* (Oxford, 1972).

6. WH, "Observations of the New Planet," unpublished paper read to the Royal Society February 18, 1802, *The Scientific Papers of Sir William Herschel*, 2 vol. (London, 1912), ed. J. L. E. Dreyer, vol. 1, pp. cix–cxi.

7. *Edinburgh Review*, 2nd issue, early 1803. The writer is anonymous but probably Brougham.

8. C. F. Gauss to H. W. M. Olbers, June 25, 1802, translation in HFA.

9. A. F. O. Alexander, *The Planet Saturn: A History of Observation, Theory and Discovery* (London, 1962).

10. WH, "Observations on the Singular Figure of the Planet Saturn," *PT*, 95 (1805), 272–80, p. 272.

11. *Chronicle*, chap. 20, p. 12.

12. WH to W. Shairp, March 9, 1794, RAS W.1/1.

13. WH to C. G. Woide, January 6, 1788, RAS W.1/1.

14. Emilio Bautista Paz et al., "Industrial Archaeology. From the 17th to the 21st Century: Reconstruction of Herschel's Telescope," *International Symposium on History of Machines and Mechanisms*, ed. Marco Ceccarelli (Dordrecht, 2004), 259–68.

15. Anthony Barrett, "A Newly-Discovered Letter from William Herschel to William Hamilton," *JRAS Canada*, 77 (1983), 167–76.

16. The story is told in *Chronicle*, 143, quoting from Charles Burney's *Memoirs of Dr Burney* (London, 1832).

17. C. Burney to WH, September 3, 1799, RAS W.1/13.B.178.

11 ▪ 1792–1822: The Torch is Handed On

1. Caroline to John, February 3, 1829, BL, Egerton 3761.

2. A family tree in the HFA shows that Mary also had a son William, who died in 1783.

3. *Chronicle*, 238. The letter is in the HFA.

4. BL, microfilm M/588(4).

5. WH, "Remarks on Mr Michell's Telescope," RAS W.7/14.

6. BL, microfilm M/588(4).

7. BL, microfilm M/588(4).

8. Marion Hardcastle, "Concerning John," HFA.

9. Ibid.

10. John Gretton, son of Dr. George Gretton, cited by Marion Hardcastle, ibid.

11. Caroline's biographical notes cited by Marion Hardcastle, ibid.

12. Archives of Eton College.

13. B. Greatheed to WH, November 2, 1800, RAS W.1/13.G.16.

14. *Chronicle*, 298.

15. John had a print of pupils "Dedicated to the Nobility and Gentry educated at Hitcham House," Marion Hardcastle, "Concerning John," HFA.

16. "Hitcham House," Alan Senior, *Hitcham and Taplow Society Newsletter*, no. 83 (2005).

17. John Herschel, "Recovery of dates," HFA.

18. For details of this period in Caroline's life, see her own records in *Memoir*, and also *Partnership*, chap. 4.

19. RAS W.6/8.

20. C. H. Marshall, "List of Subscribers of the Circulating Library [of Music], 1793–99," Bath Central Library MS 23922.

21. *Bath Chronicle*, March 7, 1799.

22. For further details of the later stages of the sweeps for nebulae, see Michael Hoskin, "Unfinished Business: William Herschel's Sweeps for Nebulae," *History of Science*, 43 (2005), 305–20.

23. WH to Caroline, July 22, 1802, HFA. For William's record of the events that followed, see RAS W.7/15.

24. Made by Noël Simon Carochez (c. 1745–1813/14). It was completed in May 1800 and a stand for it was constructed by Tremel the following year.

25. RAS W.7/15, entry for July 25, 1802.

26. John to Caroline, April 17, 1832, *Chronicle*, 381.

27. On the decline of the 40-foot, see Michael Hoskin, "Herschel's 40ft Reflector: Funding and Functions," *JHA*, 34 (2003), 1–32.

28. P. Wilson to WH, August 10, 1807, RAS W.1/13.W.166.

29. RAS W.7/10, warrant signed by King George IV, September 21, 1820.

30. Cited in *Chronicle*, 168.

31. David B. Pickering, "The Astronomical Fraternity of the World, Part III," *Popular Astronomy*, 35 (1927), 435–47, pp. 442–43.

32. The standard biography of John is Günther Buttmann, *The Shadow of the Telescope: A Biography of John Herschel* (Guildford, UK, 1974).

33. John to J. W. Whittaker, July 2, 1813, Crowe, letter 80.

34. The letters exchanged between father and son are in HFA.

35. John to C. Babbage, December 18, 1815, Crowe, letter 160.

36. John to J. W. Whittaker, September 2, 1816, Library of St. John's College, Cambridge.

37. *Chronicle*, chap. 25, p. 18.

38. *Chronicle*, chap. 25, p. 19.

12 ▪ 1822–1833: John's "Sacred Duty"

1. For further information about Caroline's life in the years following William's death, see *Memoir*, and *Partnership*, chap. 5. The many letters she wrote from Hanover to John or his wife Margaret are preserved in BL, Egerton 3761 and 3762.

2. National Archives, Kew, England, Prob 10/4640. See P. D. Hingley, "The Will of Sir William Herschel," *Astronomy & Geophysics*, 39(3) (1998), 7.

3. Herschel/M 1090, 1, Harry Ransom Library, University of Texas at Austin.

4. Caroline to Mary Herschel, April 5, 1840, BL, Egerton 3762.

5. John to Sir David Smith, September 25, 1822, Crowe, letter 801.

6. Caroline to John, May 4, 1843, BL, Egerton 3762.

7. Agnes M. Clerke, *The Herschels and Modern Astronomy* (London, 1895), 132.

8. John to Caroline, April 18, 1825, *Memoir*, 188.

9. John to Caroline, May 4–11, 1827, *Memoir*, 213.

10. Caroline to John, February 1, 1826, BL, Egerton 3761.

11. *Memoir*, 225. The gold medal is now in Girton College, Cambridge.

12. John to Caroline, May 28, 1828, *Memoir*, 227. On the rules governing Royal Society medals at the time, see Marie Boas Hall, *All Scientists Now: The Royal Society in the Nineteenth Century* (London, 1984).

13. *Chronicle*, 369.

14. John Herschel, "Account of Some Modifications Made with a 20-feet Reflecting Telescope," *Memoirs of the Astronomical Society of London*, 2 (1826), 459–97, p. 461.

15. John to C. Babbage, February 12, 1828, Crowe, letter 1676.

16. J. F. W. Herschel, "Observations of Nebulae and Clusters, Made at Slough . . . ," *PT*, 123 (1833), 359–506.

13 • 1833–1848: "The Completion of My Father's Work"

1. Mary died in January 1832 and in her will (The National Archives, Kew, England, Prob 11/1794) she left a total of £1900 to her nephews and nieces and some minor bequests to servants. Otherwise everything went to John.

2. Caroline to Margaret Herschel, December 4, 1832, BL, Egerton 3761.

3. For John's years at the Cape, see *Herschel at the Cape: Diaries and Correspondence of Sir John Herschel, 1834–1838*, ed. David S. Evans et al. (Austin, 1969). Fifty years before, William had dreamed of observing there, RAS W.4/1.5, 400.

4. John Herschel, *Results of Astronomical Observations Made . . . at the Cape of Good Hope, Being a Completion of a Telescopic Survey of the Whole of the Visible Heavens, Commenced in 1825* (London, 1847).

5. His sketches are reproduced in Brian Warner, *Cape Landscapes: Sir John Herschel's Sketches 1834–1838* (Cape Town, 2006).

6. *Bath Chronicle*, April 11, 1793.

7. The complete text of the articles is available at www.museumofhoaxes.com/moonhoax.html.

8. *Neueste Berichte vom Cap der Guten Hoffnung über Sir John Herschels höchst merkwürdige astronomische Entdeckungen den Mond und seine Bewohner betreffend* (Hamburg, 1836).

9. On the Rosse telescopes and the work done with them, see Michael Hoskin, "Rosse, Robinson, and the Resolution of the Nebulae," *JHA*, 21 (1990), 331–44.

10. *Report of the Fifteenth Meeting of the British Association for the Advancement of Science* (London, 1846), p. xxxvi.

11. Caroline to Margaret Herschel, October 3, 1844, BL, Egerton 3762.

12. The text is reproduced in *CHA*, along with an account of the circumstances of its composition.

13. G. B. Airy to John, October 13, 1846, Crowe, letter 6769.

14. Caroline to Margaret, March 1844, BL, Egerton 3762.

15. John to Caroline, July 11, 1847, *Memoir*, 342.

16. BL, microfilm M/588(4).

17. BL, microfilm M/588(4).

Bibliographic Essay

The published works of William Herschel, with one minor exception, appeared in the Royal Society's *Philosophical Transactions*. They were assembled, together with the text of the many unpublished papers that William read to the Bath Philosophical Society, by the indefatigable J. L. E. Dreyer and published in two quarto volumes with the title *The Scientific Papers of Sir William Herschel* (London, 1912). Dreyer supplies a lengthy introduction that is all the more valuable in being by an experienced astronomical observer and skilled historian who had privileged access to Herschel manuscripts.

The bulk of the scientific manuscripts of William, Caroline, and John were donated to the Royal Astronomical Society, which has made the entire collection available worldwide in the form of inexpensive CDs and DVDs. Other important holdings, especially of the papers of John, are in the library of the Royal Society. Unfortunately other family papers, mostly of secondary interest, were dispersed at auction in 1958, but microfilms of those exported from the United Kingdom are in the British Library. A small number of papers remain with the family, who are most generous in allowing access to them.

Caroline's diaries and letters are extensively cited in *Memoir and Correspondence of Caroline Herschel*, by Mrs. John Herschel, 2nd ed. (London, 1879). A granddaughter of William also drew extensively on family papers for her admirable *The Herschel Chronicle: The Life-story of William Herschel and his Sister Caroline Herschel*, ed. Constance A. Lubbock (Cambridge, 1933); this has been reprinted by the William Herschel Society, Bath, UK. Lady Lubbock was elderly when she compiled her book, and the editor of Cambridge University Press was ruthless with it, omitting extensive sections. Fortunately the original typescript is in the possession of the William Herschel Society.

Caroline wrote two autobiographies, the second when she was in her nineties. Both are incomplete but they are invaluable sources for both her life and William's: *Caroline Herschel's Autobiographies*, ed. Michael Hoskin (Cambridge, 2003).

There are many popular and semipopular discussions of William and Caroline's lives and scientific work, but *William Herschel: Explorer of the Heavens*, by

J. B. Sidgwick (London, 1953) retains its value. The inquirer after specific aspects of William's work will do well to start with Angus Armitage's methodical *William Herschel* (London, 1962).

The authoritative account of William's work as a telescope builder is J. A. Bennett, "'On the Power of Penetrating into Space': The Telescopes of William Herschel," *Journal for the History of Astronomy*, 7 (1976), 75–108. William's principal theoretical papers in astronomy are reprinted with discussion in *William Herschel and the Construction of the Heavens*, by Michael A. Hoskin (London, 1963). An account of the partnership between William and Caroline is *The Herschel Partnership: As Viewed by Caroline*, by Michael Hoskin (Cambridge, 2003). For biographical essays on William's and Caroline's parents, and on each of their siblings, see *The Herschels of Hanover*, by Michael Hoskin (Cambridge, 2007).

An exhaustive analysis of the nebulae catalogues of William and John Herschel is to be found in Wolfgang Steinicke, *Observing and Cataloguing Nebulae and Clusters: From Herschel to Dreyer's New General Catalogue* (Cambridge, 2010).

Further Reading

ALEXANDER, A. F. O.
The Planet Saturn: A History of Observation, Theory and Discovery (London, 1962).
The Planet Uranus: A History of Observation, Theory and Discovery (London, 1965).

BENNETT, J. A.
"Herschel's Scientific Apprenticeship and the Discovery of Uranus," in G. E. Hunt, ed., *Uranus and the Outer Planets* (Cambridge, 1982), 35–53.

BROWN, FRANK
Caroline Herschel as a Musician (Bath, 2000).

BUTTMANN, GÜNTHER
The Shadow of the Telescope: A Biography of John Herschel (Guildford, 1974).

CHAPMAN, ALLAN
"An Occupation for an Independent Gentleman: Astronomy in the Life of John Herschel," *Vistas in Astronomy*, 36 (1993), 71–116.

CLERKE, AGNES M.
The Herschels and Modern Astronomy (London, 1895).

CROWE, MICHAEL J.
The Extraterrestrial Life Debate, 1750–1900: The Idea of a Plurality of Worlds from Kant to Lowell (Cambridge, 1986).

EVANS, DAVID S., TERENCE J. DEEMING, B. H. EVANS, AND STEPHEN GOLDFARB
Herschel at the Cape: Diaries and Correspondence of Sir John Herschel, 1834–1838 (Cape Town, 1969).

FERGUSON, JAMES
Astronomy Explained upon Sir Isaac Newton's Principles, 2nd ed. (London, 1757).

GLYN JONES, KENNETH
The Search for the Nebulae (Chalfont St. Giles, UK, 1975).

HERSCHEL, JOHN F. W.
Results of Astronomical Observations Made During the Years 1834, 5, 6, 7, 8 at the Cape of Good Hope, Being a Completion of a Telescopic Survey of the Whole Surface of the Visible Heavens, Commenced in 1825 (London, 1847).

HINGLEY, P. D.
"The Will of Sir William Herschel," *Astronomy & Geophysics*, 39(3) (1998), 7.

HOLMES, RICHARD
The Age of Wonder (London, 2008).

HOSKIN, MICHAEL
"Alexander Herschel: The Forgotten Partner," *JHA*, 35 (2004), 387–420.
"Caroline Herschel: Assistant Astronomer or Astronomical Assistant," *History of Science*, 40 (2002), 425–44.
"Caroline Herschel as Observer," *JHA*, 36 (2005), 373–406.
"Caroline Herschel's Catalogue of Nebulae", *JHA*, 37 (2006), 251–55.
"Caroline Herschel's Revenge," *JHA*, 37 (2006), 109–10.
"George III's Purchase of Herschel Reflectors," *JHA*, 39 (2008), 121–24.
"Herschel's 40ft Reflector: Funding and Functions," *JHA*, 34 (2003), 1–32.
"Herschel's Determination of the Solar Apex," *JHA*, 11 (1980), 153–63.
"John Herschel and Astronomy: A Bicentennial Appraisal," in *John Herschel 1792–1992*, ed. B. Warner (Cape Town, 1992), 1–17.
"John Herschel's Cosmology," *JHA*, 18 (1987), 1–34.
"Nebulae, Star Clusters and the Milky Way: From Galileo to Herschel," *JHA*, 39 (2008), 363–96.
"Rosse, Robinson, and the Resolution of the Nebulae," *JHA*, 21 (1990), 331–44.
"The Leviathan of Parsonstown: Ambitions and Achievements," *JHA*, 33 (2002), 57–70.
"Unfinished Business: William Herschel's Sweeps for Nebulae," *History of Science*, 43 (2005), 305–20.
"Vocations in Conflict: William Herschel in Bath, 1766–1782," *History of Science*, 41 (2003), 315–33.
"Was William Herschel a Deserter?" *JHA*, 35 (2004), 356–58.
"William Herschel's Early Investigations of Nebulae: A Reassessment," *JHA*, 10 (1979), 165–76.

HOSKIN, MICHAEL, AND DAVID W. DEWHIRST
"William Herschel and the Prehistory of Stellar Spectroscopy," *JHA*, 37 (2006), 393–403.

HYSOM, E. J.
"Tests of the Shape of Mirrors by Herschel," *JHA*, 27 (1996), 349–52.

JAMES, KENNETH E.
"Concert Life in 18th Century Bath," PhD. diss., London University, 1987.

LESSENS, RONALD
William Herschel: Musicien Astronome (Vannes, 2004).

MAURER, ANDREAS
"A Compendium of All Known William Herschel Telescopes," *Journal of the Antique Telescope Society*, no. 14 (1998), 4–15.

PORTER, ROY
"William Herschel, Bath, and the Philosophical Society," in G. E. Hunt, ed., *Uranus and the Outer Planets* (Cambridge, 1982), 23–34.

SCHAFFER, SIMON
"'The Great Laboratories of the Universe': William Herschel on Matter Theory and Planetary Life," *JHA*, 11 (1980), 81–111.
"Uranus and the Establishment of Herschel's Astronomy," *JHA*, 12 (1981), 11–26.

SMITH, ROBERT
A Compleat System of Opticks (Cambridge, 1738).

SPAIGHT, JOHN TRACY
"'For the Good of Astronomy': The Manufacture, Sale, and Distant Use of William Herschel's Telescopes," *JHA*, 35 (2004), 45–69.

STEAVENSON, W. H.
"A Peep into Herschel's Workshop," *Transactions of the Optical Society*, 26 (1924–25), 210–37.

TURNER, A. J.
Science and Music in 18th Century Bath (Bath, 1977).

WARNER, BRIAN
"Sir John Herschel at the Cape of Good Hope," in *John Herschel 1792–1992*, ed. B. Warner (Cape Town, 1992), 19–55.

WARNER, BRIAN, AND NANCY WARNER
Maclear & Herschel: Letters & Diaries at the Cape of Good Hope 1834–1838 (Cape Town, 1984).

WHITSON, BRUCE N.
"William Herschel's 'Ecchoe Catch'," *JHA*, 39 (2008), 397–404.

WINTERBURN, EMILY
"British Library Microfilms of Herschel Materials," *JHA*, 37 (2006), 343–48.

WOODFIELD, IAN
The Celebrated Quarrel Between Thomas Linley (Senior) and William Herschel: An Episode in the Musical Life of 18th Century Bath (Bath, 1977).

Articles in many astronomy-related journals, including *Journal for the History of Astronomy* and *History of Science*, are available for download without charge from the SAO/NASA Astrophysics Data System (ADS) Web site, http://adswww.harvard.edu.

Index

Watson, William, Sr., 50, 62
Watt, James, 158
Whirlpool Nebula, 178
White, Susan, 161
Wilson, Patrick, 81, 154, 172

Windsor Castle, telescopes, 65, 109
Wollaston, William Hyde, 181
Wright, Thomas, of Durham, 100

Zach, Baron Franz Xaver von, 151